Energy Statistics

NATURAL WORLD INFORMATION GUIDE SERIES

Series Editor: Russell Shank, University Librarian, Library Administrative Office, University of California, Los Angeles

Also in this series:

RANGE SCIENCE—*Edited by John F. Vallentine and Phillip L. Sims*

The above series is part of the
GALE INFORMATION GUIDE LIBRARY

The Library consists of a number of separate series of guides covering major areas in the social sciences, humanities, and current affairs.

General Editor: Paul Wasserman, Professor and former Dean, School of Library and Information Services, University of Maryland

Managing Editor: Denise Allard Adzigian, Gale Research Company

Energy Statistics

A GUIDE TO INFORMATION SOURCES

Volume 1 in the Natural World Information Guide Series

Sarojini Balachandran

Assistant Engineering Librarian
and
Associate Professor of Library Administration
Library, University of Illinois at Urbana-Champaign

Gale Research Company
Book Tower, Detroit, Michigan 48226

Library of Congress Cataloging in Publication Data

Balachandran, Sarojini.
   Energy statistics.

   (Natural world information guide series ; v. 1)
   Includes indexes.
   1. Power resources—Statistics—Bibliography.
I. Title. II. Series.
Z5853.P83B25 [HD9502.A2] 016.33379′02′12 80-13338
ISBN 0-8103-1419-3

VITA

Sarojini Balachandran has a bachelor of science and a master of arts degree in physics from Madras University in India, where she also worked as Scientific Officer (Research & Development) at the Atomic Energy Commission. She also has a master's degree in physics from Indiana State University, Terre Haute, Indiana, and an MSLS from the University of Illinois at Urbana-Champaign, where she is currently the Assistant Engineering Librarian.

Balachandran has written TECHNICAL WRITING (1977), EMPLOYEE COMMUNICATION (1976) and REFERENCE BOOK REVIEW INDEX (1979) and is a regular contributor to the REFERENCE SERVICES REVIEW and the SERIALS REVIEW.

CONTENTS

ABBREVIATIONS

AFM	Aussenhandels verband fur Mineralol
API	American Petroleum Institute
BTU	British thermal unit
CAB	Civil Aeronautics Board (U.S.)
DOT	Department of Transportation (U.S.)
ECE	Economic Commission for Europe (United Nations)
EEC	European Economic Community
EIU	Economist Intelligence Unit
EIC	Environment Information Center
EPA	Environmental Protection Agency (U.S.)
ERDA	Energy Research and Development Administration (U.S.)
FEA	Federal Energy Administration (U.S.)
FPC	Federal Power Commission (U.S.)
GNP	Gross National Product
HMSO	Her Majesty's Stationery Office (U.K.)
HPI	Hydrocarbon processing industry
IAEA	International Atomic Energy Agency
ICC	Interstate Commerce Commission (U.S.)
LNG	Liquefied Natural Gas
LP	Liquefied Petroleum
LPG	Liquefied Petroleum Gas
MCF	Thousand Cubic Feet
NGL	Natural Gas Liquid
OAPEC	Organization of Arab Petroleum Exporting Countries

Abbreviations

OECD	Organization for Economic Cooperation and Development
OPEC	Organization of Petroleum Exporting Countries
PAD	Petroleum Administration for Defense
P & E	Production and Expansion
REA	Rural Electrification Administration (U.S.)
SMM	Sales and Marketing Management
SMSA	Standard Metropolitan Statistical Area

INTRODUCTION

While there may be many conflicting opinions as to what caused our energy crisis and a plethora of contradictory suggestions as to how best to confront it, there is absolute unanimity as to the existence of the problem. At the present time, a tremendous amount of time and money is being expended in an effort to find a lasting solution. The issues regarding the cost and probable impact of alternative energy options are being hotly debated at all levels. The general public is naturally bearing the brunt of this crisis in terms of increased fuel bills and scarcity of available supply. The subject of energy is therefore quite topical, to say the least, which explains the voluminous amount of published material relating to various aspects of the problem. It is also reflected in the numerous energy related inquiries received by academic and public libraries from people in all walks of life. A large part of such inquiries are for statistics dealing variously with the exploration, production, distribution and consumption of various fuel sources.

The recent years have seen a growth in the number of reference publications devoted exclusively to providing such statistical data. In an effort to identify and list all such publications, I had earlier prepared two brief guides entitled ENERGY STATISTICS: A GUIDE TO SOURCES and ENERGY STATISTICS: AN UPDATE published as the Council of Planning Librarians Exchange Bibliography nos. 1065 and 1247. These were received favorably and the numerous suggestions and comments I received in response to the above encouraged me to expand the project considerably. The present volume is the result of this effort.

This guide is divided into three major sections. The first contains a detailed alphabetical subject/keyword analysis of all recurring statistical data contained in some forty most used national and international energy serials. The idea behind this section is quite simple. It has been my experience as the Reference Librarian at the University of Illinois Engineering Library that most inquiries relating to energy statistics are repetitive. It is also true that almost all of them can be answered by reference to statistical serials of the type mentioned in section 2 of this book. However, it is not always easy to remember the sources of such data which are of a recurring nature. Depending on how good one's memory is, hunting for such data involves considerable time and effort. In an effort to save the valuable time of researchers, I have compiled an alphabetical subject/keyword list of specific statistical information relating to all major forms of energy. Each entry identifies the source which gives the desired information. Experience has shown that depending on the level of the patrons, most data are required either on an annual, semiannual, quarterly, monthly, weekly or daily basis. I have therefore tried to provide for most entries an indication of the specific time series in which the information is given, e.g., annual, daily, and so on.

In addition to time series information, patrons might also require geographic information, in terms of state, national, and regional energy statistics. To the extent possible, I have given entries under the name of specific state, nation, or region. In fact, if the user is looking for data on oil production in Egypt, he or she would be well advised to look directly under Egypt first. However, users will also find similar information cross referenced under OIL PRODUCTION, by country or area. Two more points need to be kept in mind for maximum use of this section. In some cases, types of data are combined in one entry, instead of two or three. This is mainly due to the fact that the keywords are interrelated and a person looking for one would normally look for the other. An example of this would be:

> Electric utilities, privately owned, number of customers, kilowatt hour sales and revenues for individual utilities. STATISTICS OF PRIVATELY OWNED UTILITIES.

> Electric utilities, publicly owned, generating stations by type of prime mover and generating capacity for individual utilities. STATISTICS OF PUBLICLY OWNED UTILITIES.

The second point which one needs to bear in mind is that, in most of the serial sources which I have analyzed, certain types of data are lumped together in one table or graph in an effort to present an overall picture of the specific situation. In such cases, it was felt that the keyword entries also be arranged such that these types of data are kept in one place. This would enable users to find all they would need in one place instead of two or more places. In all other cases, the entries relate to just one specific data relating to energy.

The second section of this book gives full bibliographic descriptions of the sources analyzed in the earlier section. The rest of the book is an annotated guide to additional sources of statistical information on individual sources of energy such as oil, petroleum, electricity, natural gas, nuclear and solar power. It is hoped that this section would help reference and acquisition librarians to develop or perfect their collection in this important area. For this purpose, a directory of publishers has been conveniently provided.

In preparing this manuscript, I received advice and encouragement from sources too numerous to mention here. My special thanks are due to Prof. Russell Shank of UCLA for his approval of my proposal and editorial guidance.

Sarojini Balachandran

Section I
KEYWORD/SUBJECT DESCRIPTORS

Abu Dhabi, crude oil prices by oil field. PLATT'S OIL PRICE HANDBOOK AND OILMANAC

Abu Dhabi, crude oil production, exports to the United States, proved crude and natural gas reserves. TWENTIETH CENTURY PETROLEUM STATISTICS

Abu Dhabi, crude oil production in barrels per day. INTERNATIONAL OIL DEVELOPMENTS

Abu Dhabi, crude oil production in barrels per day and as a percent of total production. INTERNATIONAL ECONOMIC REPORT OF THE PRESIDENT

Accident. See also Injuries

Accident frequency and severity rates of selected industries. GAS FACTS

Accident rates in gas utility industry by type of gas. GAS FACTS

Aden, refining capacity. TWENTIETH CENTURY PETROLEUM STATISTICS

Advanced gas-cooled, graphite-moderated nuclear reactors, and net electrical power generation by country. POWER REACTORS IN MEMBER STATES

Advertising expenditures of major oil companies, total and by media. NATIONAL PETROLEUM NEWS FACT BOOK

Afghanistan, proven natural gas reserves. TWENTIETH CENTURY PETROLEUM STATISTICS

Africa, capital and exploration expenditures and gross and net investments in fixed assets, by domestic petroleum industry. CAPITAL INVESTMENTS IN THE PETROLEUM INDUSTRY

Africa, crude oil production, by country, in barrels per day. INTERNATIONAL OIL DEVELOPMENTS

Africa, crude oil production, exports to the United States, proved crude and natural gas reserves, production-demand ratio, refinery capacity, refined products, and tank ship fleet. TWENTIETH CENTURY PETROLEUM STATISTICS

Africa, crude oil production in barrels per day and as a percent of production. INTERNATIONAL ECONOMIC REPORT OF THE PRESIDENT

Africa, exports of crude petroleum by country, annual. WORLD ENERGY SUPPLIES

Africa, exports of oil to selected areas, annual. BASIC PETROLEUM DATA BOOK

Africa, exports of solid fuel, by country, annual. WORLD ENERGY SUPPLIES

Air carriers, consumption of aviation gasoline and jet fuel, annual. ENERGY STATISTICS

Aircraft fuel expenses and consumption, domestic, trunk and local service carriers, annual. ENERGY STATISTICS

Airlines, jet operating expenses including fuel and oil costs by type. ENERGY STATISTICS

Alabama, crude production, daily average production per well, new wells, footage drilled, proved crude oil reserved, well completions, and oil-producing wells. TWENTIETH CENTURY PETROLEUM STATISTICS

Alabama, installed generating capacity, electricity generation by type of prime mover, and electric energy sold by type of customer. EDISON ELECTRIC INSTITUTE STATISTICAL YEARBOOK

Alabama, oil and natural gas production and reserves, exploration and development, prices and related data, annual. OIL PRODUCING INDUSTRY IN YOUR STATE

Alabama, oil companies' market share, by company. NATIONAL PETROLEUM NEWS FACT BOOK

Alabama, revenue from electricity sales to residential, commercial, industrial, and other classes of service. EDISON ELECTRIC INSTITUTE STATISTICAL YEARBOOK

Alabama, total oil and gas wells, dry holes, development, and exploratory wells drilled, by quarter. QUARTERLY REVIEW OF DRILLING STATISTICS

Alaska, crude oil prices at oil fields, by company. PLATT'S OIL PRICE HANDBOOK AND OILMANAC

Alaska, crude production, new wells, footage drilled, proved crude reserves, well completions, oil-producing wells, marketed production of natural gas, and natural gas reserves. TWENTIETH CENTURY PETROLEUM STATISTICS

Alaska, installed generating capacity, electricity generation by type of prime mover, and electric energy sold by type of customer. EDISON ELECTRIC INSTITUTE STATISTICAL YEARBOOK

Alaska, oil and natural gas production, reserves, exploration, development, prices, and related data, annual. OIL PRODUCING INDUSTRY IN YOUR STATE

Alaska, oil companies' market share, by company. NATIONAL PETROLEUM NEWS FACT BOOK

Alaska, total oil and gas wells, dry holes, development, and exploratory wells drilled, on- and offshore, by quarter. QUARTERLY REVIEW OF DRILLING STATISTICS

Albania, crude production and crude and natural gas reserves. TWENTIETH CENTURY PETROLEUM STATISTICS

Albany (N.Y.), distillates and residual fuel oils, refinery and terminal prices, high, low, and annual. PLATT'S OIL PRICE HANDBOOK AND OILMANAC

Albany (N.Y.), motor gasoline, refinery and terminal prices, regular and premium. PLATT'S OIL PRICE HANDBOOK AND OILMANAC

Algeria, crude oil production, exports to the United States, crude and natural gas reserves, refined products demand, and tank ship fleet. TWENTIETH CENTURY PETROLEUM STATISTICS

Algeria, crude oil production in barrels per day. INTERNATIONAL OIL DEVELOPMENTS

Algeria, crude oil production in barrels per day and as a percent of total production. INTERNATIONAL ECONOMIC REPORT OF THE PRESIDENT

Algeria, exports to the United States and other developed countries, total value. INTERNATIONAL OIL DEVELOPMENTS

Algeria, oil reserves and production, exports of crude oil and petroleum to the United States, annual. BASIC PETROLEUM DATA BOOK

AMOCO Production Company, crude oil prices (posted). PLATT'S OIL PRICE HANDBOOK AND OILMANAC

Anacortes (Wash.), propane refinery and terminal prices, high, low, and average. PLATT'S OIL PRICE HANDBOOK AND OILMANAC

Angola, crude production, proved reserves, refining capacity, and refined products demand. TWENTIETH CENTURY PETROLEUM STATISTICS

3

Angola, crude production, proved reserves, refining capacity, and refined products demand. TWENTIETH CENTURY PETROLEUM STATISTICS

Anthracite coal, as a percent of total domestic energy sources, annual. STANDARD AND POOR'S TRADE AND SECURITIES STATISTICS

Anthracite coal, domestic consumption, annual. COAL FACTS

Anthracite coal, domestic consumption by household, commercial, industrial, transportation, and electricity generation users, annual. GAS FACTS

Anthracite coal, domestic exports to foreign countries by country and region, volume and value, and by customs district. WORLD COAL TRADE

Anthracite coal, domestic monthly production and consumption. MONTHLY ENERGY REVIEW

Anthracite coal, domestic production, annual. COAL FACTS; GAS FACTS

Anthracite coal, domestic production, consumption, average value of mines, exports, and wholesale prices, monthly and annual. STANDARD AND POOR'S TRADE AND SECURITIES STATISTICS

Anthracite coal, domestic production and consumption from water and nuclear power, annual. BITUMINOUS COAL DATA

Anthracite coal, domestic production in BTUs. TWENTIETH CENTURY PETROLEUM STATISTICS

Anthracite coal, domestic reserves, estimates by state. WORLD COAL TRADE

Anthracite coal, forecast of selected statistical data and annual growth rate for the United States. PREDICASTS

Anthracite coal, forecast of selected statistical data and annual growth rate for the world, by region and country. WORLD CASTS

APCO Oil Corporation, crude oil prices (posted). PLATT'S OIL PRICE HANDBOOK AND OILMANAC

Aquifer storage pools in the United States. GAS FACTS

Arab and non-Arab oil imports by the United States, Japan, European community, United Kingdom, France, West Germany, and Italy. INTERNATIONAL ECONOMIC REPORT OF THE PRESIDENT

Arabian gulf, crude oil prices by oil field. PLATT'S OIL PRICE HANDBOOK AND OILMANAC

Area proved productive of oil and/or gas (acres) in each state. OIL PRODUCING INDUSTRY IN YOUR STATE

Argentina, crude oil prices by oil field. PLATT'S OIL PRICE HANDBOOK AND OILMANAC

Argentina, crude production, crude and natural gas demand, refining capacity, refined products demand, and tank ship fleet. TWENTIETH CENTURY PETROLEUM STATISTICS

Arizona, installed generating capacity, electricity generation by type of customer. EDISON ELECTRIC INSTITUTE STATISTICAL YEARBOOK

Arizona, oil and natural gas production, reserves, exploration, development, prices, and related data, annual. OIL PRODUCING INDUSTRY IN YOUR STATE

Arizona, oil companies' market share, by company. NATIONAL PETROLEUM NEWS FACT BOOK

Arizona, total oil and gas wells, dry holes, development, and exploratory wells drilled, by quarter. QUARTERLY REVIEW OF DRILLING STATISTICS

Arkansas, crude production, daily average production per well, natural gas and natural gas liquid production, natural gas and natural gas liquid reserves, proved crude reserves, and well completions. TWENTIETH CENTURY PETROLEUM STATISTICS

Arkansas, gasoline and fuel oil prices. OIL DAILY

Arkansas, installed generating capacity, electricity generation by type of prime mover, and electric energy sold, by type of customer. EDISON ELECTRIC INSTITUTE STATISTICAL YEARBOOK

Arkansas, motor gasoline refinery and terminal prices, regular and premium. PLATT'S OIL PRICE HANDBOOK AND OILMANAC

Arkansas, oil and natural gas production, reserves, exploration, development, prices, and related data, annual. OIL PRODUCING INDUSTRY IN YOUR STATE

Arkansas, oil companies' market share, by company. NATIONAL PETROLEUM NEWS FACT BOOK

Arkansas, refinery and terminal prices of distillates and residual fuel oils. PLATT'S OIL PRICE HANDBOOK AND OILMANAC

Arkansas, total oil and gas wells, dry holes, development, and exploratory wells drilled, by quarter. QUARTERLY REVIEW OF DRILLING STATISTICS

Armenia, tank ship fleet, deadweight tonnage. TWENTIETH CENTURY PETROLEUM STATISTICS

Ashland Oil, Inc., crude oil prices (posted). PLATT'S OIL PRICE HANDBOOK AND OILMANAC

Asia, crude oil production, proved crude and natural reserves, refined products demand, and tank ship tonnage. TWENTIETH CENTURY PETROLEUM STATISTICS

Asphalt, contribution to freight revenue of U.S. railroads, annual. ENERGY STATISTICS

Asphalt, domestic production, annual. STANDARD AND POOR'S TRADE AND SECURITIES STATISTICS

Asphalt, domestic production at refineries and percent refinery yield. TWENTIETH CENTURY PETROLEUM STATISTICS

Asphalt (petroleum), domestic production, monthly and annual totals. STANDARD AND POOR'S TRADE AND SECURITIES STATISTICS

Assets of gas and electric utilities in the United States. FINANCIAL STATISTICS OF PUBLIC UTILITIES; MOODY'S PUBLIC UTILITY MANUAL; OIL AND GAS CHEMICAL SERVICE; STANDARD AND POOR'S INDUSTRY SURVEYS

Atlanta (Ga.), motor gasoline, refinery and terminal prices, regular and premium. PLATT'S OIL PRICE HANDBOOK AND OILMANAC

Atlanta (Ga.), refinery and terminal prices of distillates and residual fuel oils. PLATT'S OIL PRICE HANDBOOK AND OILMANAC

Atlantic Richfield Company, crude oil prices (posted). PLATT'S OIL PRICE HANDBOOK AND OILMANAC

Auger mining, production of bituminous coal, annual. BITUMINOUS COAL DATA

Australia, crude oil production in barrels per day. INTERNATIONAL OIL DEVELOPMENTS

Australia, crude production, crude and natural gas reserves, refining capacity, refined products demand, and tank ship fleet. TWENTIETH CENTURY PETROLEUM STATISTICS

Australia, natural gas liquid production in barrels per day. INTERNATIONAL OIL DEVELOPMENTS

Australia, production, consumption, imports and exports, by function and user, of energy sources by type. STATISTICS OF ENERGY

Austria, crude production, crude and natural gas reserves, refining capacity, and refined products demand. TWENTIETH CENTURY PETROLEUM STATISTICS

Austria, production, consumption, imports and exports by direction, of crude oil and crude oil products by type, and of natural gas and liquids. QUARTERLY OIL STATISTICS

Austria, production, consumption, imports, and exports, by function and user, of energy sources by type. STATISTICS OF ENERGY

Automatic gas dryers, annual manufacturers' shipments. MOODY'S PUBLIC UTILITY MANUAL

Automobile, cost of operation by size and model, including cost of gasoline and oil. ENERGY STATISTICS

Automobile, registrations, by state. NATIONAL PETROLEUM NEWS FACT BOOK

Automobile, registrations, gasoline consumption, and average consumption per vehicle, annual. STANDARD AND POOR'S INDUSTRY SURVEYS

Automobiles (passenger), average fuel efficiency in the United States, annual. ENERGY STATISTICS

Average age of world tank ship tonnage. TWENTIETH CENTURY PETROLEUM STATISTICS

Average daily production per well, by state and PAD district. TWENTIETH CENTURY PETROLEUM STATISTICS

Average deadweight tons, world tank ship fleet. TWENTIETH CENTURY PETROLEUM STATISTICS

Average drilling cost per well (United States). TWENTIETH CENTURY PETROLEUM

Average life of oil wells. TWENTIETH CENTURY PETROLEUM STATISTICS

Average price of crude. TWENTIETH CENTURY PETROLEUM STATISTICS

Average price of natural gas. TWENTIETH CENTURY PETROLEUM STATISTICS

Average speed, world tank ship fleet. TWENTIETH CENTURY PETROLEUM STATISTICS

Aviation fuels, domestic shipments to PAD districts. ENERGY STATISTICS

Aviation fuels, domestic shipments to PAD districts, by user, daily averages. ANNUAL STATISTICAL REVIEW

Aviation fuels, estimated consumption, by state. OIL PRODUCING INDUSTRY IN YOUR STATE

Aviation fuels, world cargo prices. PLATT'S OIL PRICE HANDBOOK AND OILMANAC

Aviation gasoline, domestic production and stocks, annual. STANDARD AND POOR'S TRADE AND SECURITIES STATISTICS

Aviation gasoline, free-world percentages. INTERNATIONAL PETROLEUM

Aviation gasoline, production, imports and exports and total internal consumption in OECD member countries, annual. STATISTICS OF ENERGY

Aviation gasoline, production, trade, and total and per capita apparent consumption by country, annual. WORLD ENERGY SUPPLIES

Aviation kerosene, production, exports, imports, consumption, and supply in OECD countries, annual and quarterly. QUARTERLY OIL STATISTICS

Bahrain, crude production, crude and natural gas reserves, refining capacity, and refined products demand. TWENTIETH CENTURY PETROLEUM STATISTICS

Baltimore (Md.), distillates and residual fuel oils refinery and terminal prices, high, low, and average. PLATT'S OIL PRICE HANDBOOK AND OILMANAC

Baltimore (Md.), fuel oil prices. OIL DAILY

Baltimore (Md.), gasoline prices, premium, unleaded, and regular. OIL DAILY

Barbados, refining capacity and refined products demand. TWENTIETH CENTURY PETROLEUM STATISTICS

Baton Rouge (La.), propane refinery and terminal prices, high, low, and average. PLATT'S OIL PRICE HANDBOOK AND OILMANAC

Beehive coke plants, consumption of bituminous coal, annual. BITUMINOUS COAL DATA

Belgium, production, consumption, and exports and imports by direction, of crude oil and crude oil products by type, and of natural gas and liquids. QUARTERLY OIL STATISTICS

Belgium, production, consumption, exports, and imports, by function and user, of energy sources by type. STATISTICS OF ENERGY

Belgium, refining capacity, refined products demand, and tank ship fleet. TWENTIETH CENTURY PETROLEUM STATISTICS

Belgium-Luxembourg, estimated imports of crude oil and refined products from Arab and non-Arab countries by original crude source. INTERNATIONAL OIL DEVELOPMENTS

8

Belton (S.C.), motor gasoline refinery and terminal prices, regular and premium. PLATT'S OIL PRICE HANDBOOK AND OILMANAC

Belton (S.C.), refinery and terminal prices of distillates and residual fuel oils. PLATT'S OIL PRICE HANDBOOK AND OILMANAC

Benzol, domestic production. TWENTIETH CENTURY PETROLEUM STATISTICS

Benzol, domestic production, annual. STANDARD AND POOR'S TRADE AND SECURITIES STATISTICS

Birmingham (Ala.), motor gasoline refinery and terminal prices, regular and premium. PLATT'S OIL PRICE HANDBOOK AND OILMANAC

Bitumen and road oil, world production by country, annual. WORLD ENERGY SUPPLIES

Bituminous coal. See also Coal

Bituminous coal, amount cleaned by wet and pneumatic methods. BITUMINOUS COAL DATA

Bituminous coal, average tons per car load, by selected railroads. COAL TRAFFIC ANNUAL

Bituminous coal, average value at mines by type of mine, by district, annual. BITUMINOUS COAL DATA

Bituminous coal, average value of U.S. exports. WORLD COAL TRADE

Bituminous coal, average value per ton, F.O.B. mines, by state. COAL FACTS

Bituminous coal, cleaning plant operations and coal cleaned. BITUMINOUS COAL DATA

Bituminous coal, consumption by cement mills in the United States. COAL FACTS

Bituminous coal, consumption by Electric power utilities in the United States. COAL FACTS

Bituminous coal, consumption by electric utilities, coke ovens, steel mills, manufacturing companies, and retail sales, annual. STANDARD AND POOR'S TRADE AND SECURITIES STATISTICS

Bituminous coal, consumption by individual electric utilities, by origin state. COAL TRAFFIC ANNUAL

Bituminous coal, consumption by railroads in the United States. COAL FACTS

Bituminous coal, consumption by steel and rolling mills in the United States. COAL FACTS

Bituminous coal, consumption for making coking coal in the United States. COAL FACTS

Bituminous coal, distribution by region and state of destination. COAL FACTS

Bituminous coal, domestic consumption, yearly and monthly, by consumer class, annual. BITUMINOUS COAL DATA

Bituminous coal, domestic consumption and exports, annual. COAL FACTS

Bituminous coal, domestic exports by country of destination and by continental group. BITUMINOUS COAL DATA

Bituminous coal, domestic exports by customs district. WORLD COAL TRADE

Bituminous coal, domestic inland water movements by shipping and receiving areas and tonnage handled. COAL TRAFFIC ANNUAL

Bituminous coal, domestic production, by state. COAL FACTS

Bituminous coal, domestic production, by state and all types of mines. COAL FACTS

Bituminous coal, domestic production, by state and district. BITUMINOUS COAL DATA

Bituminous coal, domestic production, by type of mine and by size of mine output, annual. BITUMINOUS COAL DATA

Bituminous coal, domestic production, monthly and weekly, and annual peak week production. BITUMINOUS COAL DATA

Bituminous coal, domestic production and consumption from water and nuclear power, annual. BITUMINOUS COAL DATA

Bituminous coal, domestic production and consumption in BTUs.

Bituminous coal, domestic production and stocks, annual. STANDARD AND POOR'S TRADE AND SECURITIES STATISTICS

Bituminous coal, domestic production by district. COAL FACTS

Bituminous coal, domestic production by method of mining. COAL FACTS

Bituminous coal, domestic production from consumer-owned (captive) mines. COAL FACTS

Bituminous coal, domestic production not sold in open market, by state. BITUMINOUS COAL DATA

Bituminous coal, dumpings at tidewater, by railroad and port. COAL TRAFFIC ANNUAL

Bituminous coal, final mode of transportation to plant, by individual electric utility, by state. COAL TRAFFIC ANNUAL

Bituminous coal, handled by class I railroads and revenue received, by individual railroad. COAL TRAFFIC ANNUAL

Bituminous coal, loaded at mine for shipment by rail. COAL FACTS

Bituminous coal, loaded at mine for shipment by water. COAL FACTS

Bituminous coal, loaded into vessels at Lake Erie and Lake Ontario ports by origin district. COAL TRAFFIC ANNUAL

Bituminous coal, mode of transportation to individual electric utilities, by state. COAL TRAFFIC ANNUAL

Bituminous coal, mode of transportation to U.S. destinations. COAL TRAFFIC ANNUAL

Bituminous coal, moves from mines. COAL FACTS

Bituminous coal, price indicators. BITUMINOUS COAL DATA

Bituminous coal, prices, average value at mines, and wholesale price, monthly and annual averages. STANDARD AND POOR'S TRADE AND SECURITIES STATISTICS

Bituminous coal, shipments, by region and state and by customer. COAL FACTS

Bituminous coal, shipments by motor vehicle. COAL FACTS

Bituminous coal, shipments from Lake Ontario and Lake Michigan ports. COAL TRAFFIC ANNUAL

Bituminous coal, shipments from the mine, by method of transportation. BITUMINOUS COAL DATA

Bituminous coal, shipments to all destinations in the United States, Canada, and Mexico, by method of movement and by consumer use. COAL FACTS

Bituminous coal, shipments to Canada and U.S. ports from Lake Erie, by loading port. COAL TRAFFIC ANNUAL

Bituminous coal, stocks and days supply in hands of consumer plants and yards, by type of consumer. BITUMINOUS COAL DATA

Bituminous coal, strip and auger mining production, by state. BITUMINOUS COAL DATA

Bituminous coal, transportation by U.S. rail. COAL FACTS

Bituminous coal, truck shipments, by origin, state, and destination state. COAL TRAFFIC ANNUAL

Bituminous coal, underground and surface production, by state. COAL FACTS

Bituminous coal, underground production, by state. BITUMINOUS COAL DATA

Bituminous coal, unit train movements, by state. COAL TRAFFIC ANNUAL

Bituminous coal and lignite. See also Coal; Lignite

Bituminous coal and lignite, domestic consumption, annual. COAL FACTS

Bituminous coal and lignite, domestic consumption, production, exports, and stocks. MONTHLY ENERGY REVIEW

Bituminous coal and lignite, domestic consumption and supply, daily averages. ANNUAL STATISTICAL REVIEW

Bituminous coal and lignite, domestic consumption by household, commercial, industrial, transportation, and electricity generation users, annual. GAS FACTS

Bituminous coal and lignite, domestic production. COAL FACTS; GAS FACTS

Bituminous coal and lignite, domestic production by state, size of output, and type of mining. ENERGY STATISTICS

Bituminous coal and lignite, estimated reserves by state. GAS FACTS

Bituminous coal and lignite, forecast of selected statistical data and annual growth rates for the United States. PREDICASTS

Bituminous coal and lignite, forecast of selected statistical data and annual growth rates for the world, by region and country. WORLD CASTS

Bituminous coal and lignite, world production by region and by country. WORLD ENERGY SUPPLIES

Bituminous coal and lignite mines, total number in the United States by state. COAL FACTS

Bituminous coal and other energy sources, domestic utilization. COAL FACTS

Bituminous coal as a percent of total U.S. energy sources, annual. STANDARD AND POOR'S TRADE AND SECURITIES STATISTICS

Bituminous coal mines, average number of men employed, by state. COAL FACTS

Bituminous coal mines, fifty biggest mines by rank. COAL FACTS

Bituminous coal mines, output per man per day by state and by district. BITUMINOUS COAL DATA

Bituminous coal mining industry, average number of days worked. COAL FACTS

Bituminous coal mining industry, domestic production, annual. COAL FACTS

Bituminous coal mining industry, growth in the United States, annual. COAL FACTS

Bituminous coal mining industry, hourly wages, weekly earnings, and hours worked, monthly and annual averages. STANDARD AND POOR'S TRADE AND SECURITIES STATISTICS

Bituminous coal mining industry, net income and federal taxes of mining corporations, annual. BITUMINOUS COAL DATA

Bituminous coal mining industry, net tons per man. COAL FACTS

Bituminous coal mining industry, number of mines. COAL FACTS

Bituminous coal mining industry, percent of total production, annual. COAL FACTS

Bituminous coal mining industry, percent of underground production. COAL FACTS

Bituminous coal mining industry, total employment. COAL FACTS

Boilers, manufacturers' shipments, annual. GAS FACTS; MOODY'S PUBLIC UTILITY MANUAL

Boiling light-water-cooled and moderated reactors, description and net electrical power generation, by country. POWER REACTORS IN MEMBER STATES

Bolivia, crude production, crude and natural gas reserves, refining capacity, and refined products demand. TWENTIETH CENTURY PETROLEUM STATISTICS

Bond issues of the gas utility industry, by type of company, cost, and yield. GAS FACTS

Bond issues of utilities, description, interest date, amount, rating, price, and yield of individual issues. MOODY'S PUBLIC UTILITY MANUAL

Boston (Mass.), distillate and residual fuel oil, cargo prices, high, low, and average. PLATT'S OIL PRICE HANDBOOK AND OILMANAC

Boston (Mass.), fuel oil prices. OIL DAILY

Boston (Mass.), gasoline prices, premium, unleaded, and regular. OIL DAILY

Boston (Mass.), motor gasoline refinery and terminal prices, regular and premium. PLATT'S OIL PRICE HANDBOOK AND OILMANAC

Brazil, crude production, crude and natural gas reserves, refining capacity, refined products demand, and tank ship fleet. TWENTIETH CENTURY PETRO-LEUM STATISTICS

British Columbia, crude oil prices, by oil field. PLATT'S OIL PRICE HAND-BOOK AND OILMANAC

Brown coal. See Lignite and brown coal

Buffalo (N.Y.), refinery and terminal prices of distillates and residual fuel oils. PLATT'S OIL PRICE HANDBOOK AND OILMANAC

Bulgaria, crude production, proven crude and natural gas reserves, and refining capacity. TWENTIETH CENTURY PETROLEUM STATISTICS

Bulgaria, oil and natural gas production, consumption, and trade. INTERNA-TIONAL OIL DEVELOPMENTS

Bulk plants operated by oil companies, total number. NATIONAL PETROLEUM NEWS FACT BOOK

Bunker "C" fuel oil, retail prices, including duty and taxes, by country. INTERNATIONAL PETROLEUM

Bunker fuel oil, prices for Gulf Coast cargoes, high, low, and average. PLATT'S OIL PRICE HANDBOOK AND OILMANAC

Burma, crude production, proven crude and natural gas reserves and refining capacity. TWENTIETH CENTURY PETROLEUM STATISTICS

Buses, domestic, total registrations and vehicle miles, average miles driven per travel and average miles per gallon, annual. BASIC PETROLEUM DATA BOOK

California, crude oil prices, monthly and annual averages. STANDARD AND POOR'S TRADE AND SECURITIES STATISTICS

California, crude production, daily average production per well, new wells drilled, natural gas and natural gas liquids production, crude reserves, and well completions. TWENTIETH CENTURY PETROLEUM STATISTICS

California, installed generating capacity, electricity generation by type of prime mover, and electric energy sold by type of customer. EDISON ELECTRIC INSTITUTE STATISTICAL YEARBOOK

California, oil and natural gas production, reserves, exploration, development, prices, and related data, annual. OIL PRODUCING INDUSTRY IN YOUR STATE

California, oil companies' market share, by company. NATIONAL PETROLEUM NEWS FACT BOOK

California, total oil and gas wells, dry holes, and development and exploratory wells drilled, by district, by quarter. QUARTERLY REVIEW OF DRILLING STATISTICS

Canada, capital and exploration expenditures and gross and net investment in fixed assets by domestic petroleum industry. CAPITAL INVESTMENTS IN THE PETROLEUM INDUSTRY

Canada, crude oil prices by oil field, by province. PLATT'S OIL PRICE HANDBOOK AND OILMANAC

Canada, crude oil production in barrels per day. INTERNATIONAL OIL DEVELOPMENTS

Canada, crude oil production in barrels per day and as a percent of total production. INTERNATIONAL ECONOMIC REPORT OF THE PRESIDENT

Canada, crude production, exports to the United States, proven crude and natural gas reserves, refining capacity, refined products demand, and tank ships. TWENTIETH CENTURY PETROLEUM STATISTICS

Canada, estimated imports of crude oil and refined products from Arab and non-Arab countries by original crude source. INTERNATIONAL OIL DEVELOPMENTS

Canada, imports and exports of crude oil and crude oil products. INTERNATIONAL OIL DEVELOPMENTS

Canada, imports of U.S. coal by method of transportation, by province. WORLD COAL TRADE

Canada, natural gas liquids production in barrels per day. INTERNATIONAL OIL DEVELOPMENTS

Canada, primary energy production, annual. INTERNATIONAL ECONOMIC REPORT OF THE PRESIDENT

Canada, production, consumption, and exports and imports by direction, of crude oil and crude oil products by type, and of natural gas and liquids. QUARTERLY OIL STATISTICS

Canada, production, consumption, exports, and imports, by function and user, of energy sources by type. STATISTICS OF ENERGY

Canadian oil companies, retail distribution outlets, total number. NATIONAL PETROLEUM NEWS FACT BOOK

Canary Islands, refining capacity. TWENTIETH CENTURY PETROLEUM STATISTICS.

Capital, new, average price and Moody's composite average yields, monthly, and annual. MOODY'S PUBLIC UTILITY MANUAL

Capital expenditures of domestic oil companies, by function. NATIONAL PETROLEUM NEWS FACT BOOK

Capital expenditures of domestic oil companies, by function. FINANCIAL ANALYSIS OF A GROUP OF PETROLEUM COMPANIES

Capital expenditures of the domestic petroleum industry by type, annual. BASIC PETROLEUM DATA BOOK

Capital expenditures of world petroleum industry, by select countries. CAPITAL INVESTMENTS IN THE PETROLEUM INDUSTRY

Capitalization and capitalization ratios of investor-owned electric utilities. EDISON ELECTRIC INSTITUTE STATISTICAL YEARBOOK

Capitalization of gas and electric utilities in the United States. FINANCIAL STATISTICS OF PUBLIC UTILITIES; MOODY'S PUBLIC UTILITY MANUAL; OIL AND GAS CHEMICAL SERVICE; STANDARD AND POOR'S INDUSTRY SURVEYS

Captive mines, production of bituminous coal, annual. BITUMINOUS COAL DATA

Carbon black plants' consumption of natural gas in the United States, annual. GAS FACTS

Carbonized coal. See Coal, carbonized

Caribbean, exports of crude petroleum to other countries by country, annual. WORLD ENERGY SUPPLIES

Car loads, average tones of coal per load by selected railraods. COAL TRAFFIC ANNUAL

Carolinas, gasoline and fuel oil prices. OIL DAILY

Cement mills, annual consumption of bituminous coal. BITUMINOUS COAL DATA

Charleston (S.C.), distillates and residual fuel oils refinery and terminal prices, high, low, and average. PLATT'S OIL PRICE HANDBOOK AND OILMANAC

Charleston (S.C.), motor gasoline refinery and terminal prices, regular and premium. PLATT'S OIL PRICE HANDBOOK AND OILMANAC

Charlotte, (N.C.), motor gasoline and refinery and terminal prices, regular and premium. PLATT'S OIL PRICE HANDBOOK AND OILMANAC

Charlotte (N.C.), refinery and terminal prices of distillates and residual fuel oils. PLATT'S OIL PRICE HANDBOOK AND OILMANAC

Chemical plants, capital and exploration expenditures and gross and net fixed investments, by world petroleum industry, by select countries. CAPITAL IN-VESTMENTS IN THE PETROLEUM INDUSTRY

Chevron Oil Company, crude oil prices (posted). PLATT'S OIL PRICE HAND-BOOK AND OILMANAC

Chicago (Ill.), gasoline and fuel oil prices. OIL DAILY

Chicago (Ill.), kerosene prices. OIL DAILY

Chicago (Ill.), motor gasoline refinery and terminal prices, regular and premium. PLATT'S OIL PRICE HANDBOOK AND OILMANAC

Chicago (Ill.), propane refinery and terminal prices, high, low, and average. PLATT'S OIL PRICE HANDBOOK AND OILMANAC

Chicago (Ill.), refinery and terminal prices of distillates and residual fuel oils. PLATT'S OIL PRICE HANDBOOK AND OILMANAC

China, crude oil production in barrels per day. INTERNATIONAL OIL DE-VELOPMENTS

Chile, crude production, proven crude and natural gas reserves, refinery capacity, refined products, demand and tank ship fleet. TWENTIETH CENTURY PETROLEUM STATISTICS

China, crude production, proven crude and natural gas reserves, refined prod-ucts demand, refinery capacity, and tank ship fleet. TWENTIETH CENTURY PETROLEUM STATISTICS

China, natural gas liquids production in barrels per day. INTERNATIONAL OIL DEVELOPMENTS

China, primary energy production, annual. INTERNATIONAL ECONOMIC REPORT OF THE PRESIDENT

Cities and towns served by gas and electric utilities in the United States. FINANCIAL STATISTICS OF PUBLIC UTILITIES; MOODY'S PUBLIC UTILITY MANUAL

Cities Service Oil Company, crude oil prices (posted). PLATT'S OIL PRICE HANDBOOK AND OILMANAC

Cleaning plant operations, by type of mine. BITUMINOUS COAL DATA

Clothes dryers, gas, automatic; annual manufacturers' shipment. MOODY'S PUBLIC UTILITY MANUAL

Coal. See also Bituminous coal; Bituminous coal and lignite; Lignite

Coal, anthracite, bituminous, and lignite; domestic consumption. MONTHLY ENERGY REVIEW

Coal, anthracite, bituminous, and lignite; domestic consumption and production in BTUs. TWENTIETH CENTURY PETROLEUM STATISTICS

Coal, anthracite, bituminous, and lignite; domestic consumption and supply. ENERGY STATISTICS

Coal, as a percent of total U.S. energy sources, annual. STANDARD AND POOR'S TRADE AND SCURITIES STATISTICS

Coal, brown and lignite, production, imports, exports, and consumption by function and by user, and consumption for transformation in OECD member countries, annual. STATISTICS OF ENERGY

Coal, carbonized and value at coke plants. BITUMINOUS COAL DATA

Coal, demand by sector, fuel versus nonfuel use, annual. BASIC PETROLEUM DATA BOOK

Coal, demand for input in domestic industrial, household, commercial, and transportation sectors, annual. BASIC PETROLEUM DATA BOOK

Coal, domestic consumption through the year 2000, by consuming sector and as a percent of energy input. BASIC PETROLEUM DATA BOOK

Coal, domestic exports to foreign countries, by county and region, volume and value, annual. WORLD COAL TRADE

Coal, domestic railroad dumpings for export. WORLD COAL TRADE

Coal, estimated domestic reserves, by state. WORLD COAL TRADE; ENERGY STATISTICS

Coal, estimated world resources. COAL FACTS

Coal, Federal Reserve Board indexes of industrial production, monthly and annual averages. STANDARD AND POOR'S TRADE AND SECURITIES STATISTICS

Coal, forecast of selected statistical data and annual growth rate for the United States. PREDICASTS

Coal, forecast of selected statistical data and annual growth rate for the world, by region and country. WORLD CASTS

Coal, import and export trade by direction, for selected countries. WORLD COAL TRADE

Coal, origin district for domestic export from Atlantic coast ports. WORLD COAL TRADE

Coal, production, imports, and exports for selected countries. WORLD COAL TRADE

Coal, production, volume, and as a percent of total U.S. energy production, annual. BASIC PETROLEUM DATA BOOK

Coal, production by leading producers and percent of world total. WORLD COAL TRADE

Coal, strippable domestic reserves. COAL FACTS

Coal, world consumption by source, through 1990, and as a percent of total energy consumption. BASIC PETROLEUM DATA BOOK

Coal, world production by country and region. WORLD COAL TRADE

Coal briquettes (brown), production, imports, exports, and consumption by function and by user, and consumption for transformation in OECD member countries, annual. STATISTICS OF ENERGY

Coal consumption by steam-electric plants, by region. COAL FACTS

Coal consumption for electricity generation, annual. EDISON ELECTRIC INSTITUTE STATISTICAL YEARBOOK

Coal consumption for electricity generation by individual utilities and cost of coal per ton. STEAM ELECTRIC PLANT FACTORS

Coal cutting machines at underground bituminous coal mines, by state. BITUMINOUS COAL DATA

Coal dumpings by railroads at Atlantic ports for export. WORLD COAL TRADE

Coal-generated electricity, total, annual and by state. EDISON ELECTRIC INSTITUTE STATISTICAL YEARBOOK

Coal (hard), production, imports, exports, consumption by function and by user, and consumption for transformation in OECD member countries, annual. STATISTICS OF ENERGY

Coal (Hard), production in selected countries, annual. INTERNATIONAL ECONOMIC REPORT OF THE PRESIDENT

Coal industry, average hourly earnings compared with other industries. COAL FACTS

Coal mines, number of active mines by type, by state, and by district. BITUMINOUS COAL DATA

Coal mines, output per man per day in the United States and Europe. COAL FACTS; WORLD COAL TRADE

Coal reserve base, demonstrated, by method of mining in the United States. COAL FACTS

Coal reserves, domestic. COAL FACTS

Coal reserves, domestic, by rank, by sulfur content, and by state. COAL FACTS

Coal reserves, domestic, by state, by type. BITUMINOUS COAL DATA

Coal reserves of the world, estimates by country and region. WORLD COAL TRADE

Coal resources, domestic and imported, of the United States, Japan, European community, United Kingdom, France, West Germany, and Italy. INTERNATIONAL ECONOMIC REPORT OF THE PRESIDENT

Coal revenue car loadings by leading coal carriers. COAL TRAFFIC ANNUAL

Coal used for metallurgical purposes by importers from the United States, by country and area. WORLD COAL TRADE

Coal utilization at steam electric plants, trends in efficiency. STEAM ELECTRIC PLANT FACTORS

Coke, beehive and by-product, production, and wholesale price index. STANDARD AND POOR'S TRADE AND SECURITIES STATISTICS

Coke, consumption (apparent), stocks, distribution, and receipts from commercial sales, annual. BITUMINOUS COAL DATA

Coke, consumption in blast furnaces, annual. BITUMINOUS COAL DATA

Coke, domestic production, annual. STANDARD AND POOR'S TRADE AND SECURITIES STATISTICS

Coke, domestic production at refineries and percent of refinery yield. TWENTIETH CENTURY PETROLEUM STATISTICS

Coke, oven and beehive, production and value by producer source, by state. BITUMINOUS COAL DATA

Coke oven coke, production, imports, exports, consumption by function and by user, and consumption for transformation in OECD member countries, annual. STATISTICS OF ENERGY

Coke oven coke, production, world and regional, annual. WORLD ENERGY
SUPPLIES

Colorado, crude production, daily average production per well, natural gas
and natural gas liquids production, proved crude and natural gas reserves, and
well completions. TWENTIETH CENTURY PETROLEUM STATISTICS

Colorado, installed generating capacity; electricity generation, by type of
prime mover; and electric energy sold, by type of customer. EDISON ELECTRIC
INSTITUTE STATISTICAL YEARBOOK

Colorado, oil and natural gas production and reserves, exploration, development,
prices, and related data, annual. OIL PRODUCING INDUSTRY IN YOUR
STATE

Colorado, oil companies' market share, by company. NATIONAL PETROLEUM
NEWS FACT BOOK

Colorado, total oil and gas wells, dry holes, development, and exploratory
wells drilled, by quarter. QUARTERLY REVIEW OF DRILLING STATISTICS

Columbia, crude production, exports to the United States, proven crude and
natural gas reserves, refining capacity, refined products demand, and tank
ship fleet. TWENTIETH CENTURY PETROLEUM STATISTICS

Combustion turbine plants, reported new capability addition, by type of
ownership. MOODY'S PUBLIC UTILITY MANUAL

Commercial customers, electric power sales, annual. STANDARD AND POOR'S
INDUSTRY SURVEYS

Commercial customers, electric power sales and revenue, annual. EDISON
ELECTRIC INSTITUTE STATISTICAL YEARBOOK

Commercial energy, consumption in the United States. MONTHLY ENERGY
REVIEW

Commercial energy, world demand by type of energy, annual. STANDARD
AND POOR'S INDUSTRY SURVEYS

Commercial sector, energy consumption in the United States, annual. BASIC
PETROLEUM DATA BOOK

Communist bloc. See also under specific Communist countries

Communist bloc, export of solid fuel to other countries by country, annual.
WORLD ENERGY SUPPLIES

Compression ratios for U.S. passenger cars, annual. NATIONAL PETROLEUM
NEWS FACT BOOK

Condensate wells, new completions in the United States. GAS FACTS

Congo, crude production, proved crude reserves, and refined products demand. TWENTIETH CENTURY PETROLEUM STATISTICS

Connecticut, installed generating capacity, electricity generation by type of prime mover; and electric energy sold by type of customer. EDISON ELECTRIC INSTITUTE STATISTICAL YEARBOOK

Connecticut, oil companies' market shares, by company. NATIONAL PETROLEUM NEWS FACT BOOK

Construction cost trends in electric light and power industry, by region. EDISON ELECTRIC INSTITUTE STATISTICAL YEARBOOK; MOODY'S PUBLIC UTILITY MANUAL

Construction expenditures by electric utilities, total, annual. MOODY'S PUBLIC UTILITY MANUAL

Construction expenditures of gas utilities by type of service, annual. MOODY'S PUBLIC UTILITY MANUAL

Consumer-owned mines, production of bituminous coal, annual. BITUMINOUS COAL DATA

Consumer price index and residential electricity. EDISON ELECTRIC INSTITUTE STATISTICAL YEARBOOK

Consumer price index for coal, electricity, fuel oil, and gas. GAS FACTS

Consumer price index for electricity and gas, annual. EDISON ELECTRIC INSTITUTE STATISTICAL YEARBOOK

Continental Oil Company, crude oil prices (posted). PLATT'S OIL PRICE HANDBOOK AND OILMANAC

Continuous mining machine production at underground bituminous coal mines, by state. BITUMINOUS COAL DATA

Contract drilling companies, estimated percentage sales to the petroleum industry and related financial data. STANDARD AND POOR'S INDUSTRY SURVEYS

Conventional steam-electric utilities, electricity generation by type of ownership, annual. EDISON ELECTRIC INSTITUTE STATISTICAL YEARBOOK

Conventional steam-electric utilities, installed generating capacity by ownership, annual. EDISON ELECTRIC INSTITUTE STATISTICAL YEARBOOK

Conventional steam-generating plants, kilowatts generated, and as a percent of total production, annual. STANDARD AND POOR'S INDUSTRY SURVEYS

Conversion burners, manufacturer's shipments, annual. GAS FACTS; MOODY'S PUBLIC UTILITY MANUAL

Conversion efficiency of energy in the United States, annual. BASIC PETRO-LEUM DATA BOOK

Conversion factors, fuel and power. COAL FACTS

Conveyors, gathering and haulage, at underground bituminous coal mines. BITUMINOUS COAL DATA

Cooperative electric utilities, installed generating capacity, annual. EDISON ELECTRIC INSTITUTE STATISTICAL YEARBOOK

Cost. See also Construction cost; Wells, cost of drilling

Cost of drilling and equipping dry holes, gas wells, offshore wells, and oil wells. TWENTIETH CENTURY PETROLEUM STATISTICS

Cost of drilling wells, average depth, cost per foot, and index of cost per foot. OIL AND GAS CHEMICAL SERVICE

Counties with oil and/or gas production in each state, total number. OIL PRODUCING INDUSTRY IN YOUR STATE

Crude oil, average annual wellhead price In the United States, in current constant dollars. BASIC PETROLEUM DATA BOOK

Crude oil, average daily production per well, by state. OIL PRODUCING INDUSTRY IN YOUR STATE

Crude oil, average field price per barrel, by state. OIL PRODUCING IN-DUSTRY IN YOUR STATE

Crude oil, average production in barrels daily, by state. OIL PRODUCING INDUSTRY IN YOUR STATE

Crude oil, average wellhead value per barrel, by state. BASIC PETROLEUM DATA BOOK

Crude oil, capital and exploration expenditures and gross and net investment by the world petroleum industry, by select countries. CAPITAL INVESTMENTS IN THE PETROLEUM INDUSTRY

Crude oil, changes in estimated proved recoverable reserves, by state. GAS FACTS

Crude oil, contribution to freight revenue of U.S. railroads, annual. ENERGY STATISTICS

Crude oil, conversion factors. BP STATISTICAL REVIEW OF THE WORLD OIL INDUSTRY

Crude oil, declining domestic supply, annual. BASIC PETROLEUM DATA BOOK

Crude oil, dollar value at the wellhead, annual. BASIC PETROLEUM DATA BOOK

Crude oil, domestic consumption, daily averages. TWENTIETH CENTURY PETROLEUM STATISTICS

Crude oil, domestic demand by area (excluding Sino-Soviet). INTERNATION-AL PETROLEUM

Crude oil, domestic direct consumption, annual. STANDARD AND POOR'S TRADE AND SECURITIES STATISTICS

Crude oil, domestic exports, by country of destination, by averages. ANNUAL STATISTICAL REVIEW

Crude oil, domestic imports. BASIC PETROLEUM DATA BOOK

Crude oil, domestic imports, annual, historical data. OIL PRODUCING INDUSTRY IN YOUR STATE

Crude oil, domestic imports, by area and country of origin, annual, and by PAD district. BASIC PETROLEUM DATA BOOK

Crude oil, domestic imports, by country of origin, daily averages. ANNUAL STATISTICAL REVIEW

Crude oil, domestic imports, exports, and net imports, annual. INTERNA-TIONAL ECONOMIC REPORT OF THE PRESIDENT

Crude oil, domestic imports, total and as a percent of domestic demand, annual. STANDARD AND POOR'S TRADE AND SECURITIES STATISTICS

Crude oil, domestic imports, total and as a percent of domestic demand, annual. STANDARD AND POOR'S INDUSTRY SURVEYS

Crude oil, declining domestic supply, annual. BASIC PETROLEUM DATA BOOK

Crude oil, dollar value at the wellhead, annual. BASIC PETROLEUM DATA BOOK

Crude oil, domestic consumption, daily averages. TWENTIETH CENTURY PETROLEUM STATISTICS

Crude oil, domestic demand by area (excluding Sino-Soviet). INTERNATIONAL PETROLEUM

Crude oil, domestic direct consumption, annual. STANDARD AND POOR's TRADE AND SECURITIES STATISTICS

Crude oil, domestic exports, by country of destination, by averages. ANNUAL STATISTICAL REVIEW

Crude oil, domestic imports. BASIC PETROLEUM DATA BOOK

Crude oil, domestic imports, annual, historical data. OIL PRODUCING IN-DUSTRY IN YOUR STATE

Crude oil, domestic imports, by area and country of origin, annual, and by PAD district. BASIC PETROLEUM DATA BOOK

Crude oil, domestic imports, by country of origin, daily averages. ANNUAL STATISTICAL REVIEW

Crude oil, domestic imports, exports, and net imports, annual. INTERNATION-AL ECONOMIC REPORT OF THE PRESIDENT

Crude oil, domestic imports, monthly and annual. STANDARD AND POOR'S TRADE AND SECURITIES STATISTICS

Crude oil, domestic imports, total and as a percent of domestic demand, annual. STANDARD AND POOR'S INDUSTRY SURVEYS

Crude oil, domestic production. See also Crude oil production

Crude oil, domestic production, annual, historical data. OIL PRODUCING INDUSTRY IN YOUR STATE

Crude oil, domestic production, by PAD district. OIL PRODUCING INDUSTRY IN YOUR STATE

Crude oil, domestic production, by state. OIL AND GAS CHEMICAL SER-VICE

Crude oil, domestic production, by state and ranked field; number of wells, year of discovery, annual production, and remaining reserves. OIL AND GAS CHEMICAL SERVICE

Crude oil, domestic production, by stripper well, percentages, by state. OIL PRODUCING INDUSTRY IN YOUR STATE

Crude oil, domestic production, demand, imports, and exports, annual. STANDARD AND POOR'S INDUSTRY SURVEYS

Crude oil, domestic production, imports, stocks, crude input to refineries. MONTHLY ENERGY REVIEW

Crude oil, domestic production, ranking by company and daily production. OIL AND GAS CHEMICAL SERVICE

Crude oil, domestic production and average production per well per day, by state, annual. BASIC PETROLEUM DATA BOOK

Crude oil, domestic production in year of largest production (barrels), by state. OIL PRODUCING INDUSTRY IN YOUR STATE

Crude oil, domestic production (net), by oil producer, and ranking by company. OIL AND GAS CHEMICAL SERVICE

Crude oil, domestic production, world production, domestic imports, exports, consumption, and year-end stocks, annual. STANDARD AND POOR'S INDUSTRY SURVEY

Crude oil, domestic productive capacity, daily estimates. RESERVES OF CRUDE OIL, NATURAL GAS LIQUID AND NATURAL GAS

Crude oil, domestic recoverable reserves, by state. BITUMINOUS COAL DATA

Crude oil, domestic reserves. See also Crude oil reserves

Crude oil, domestic reserves, discoveries, production, and indicated reserves, annual. STANDARD AND POOR'S TRADE AND SECURITIES STATISTICS

Crude oil, domestic reserves, proved, annual. ANNUAL STATISTICAL REVIEW

Crude oil, domestic reserves, proved, annual, historical data. OIL PRODUCING INDUSTRY IN YOUR STATE

Crude oil, domestic reserves, proved estimates, annual. ENERGY STATISTICS

Crude oil, domestic reserves added, annual, historical data. OIL PRODUCING INDUSTRY IN YOUR STATE

Crude oil, domestic supply and demand, annual. BASIC PETROLEUM DATA BOOK

Crude oil, estimated domestic reserves, by state. OIL AND GAS CHEMICAL SERVICE

Crude oil, estimated domestic reserves, total, and annual proved estimates RESERVES OF CRUDE OIL, NATURAL GAS LIQUID AND NATURAL GAS

Crude oil, estimated original oil in place, and ultimate recovery of crude oil in the United States, by reservoir lithology. RESERVES OF CRUDE OIL, NATURAL GAS

Crude oil, estimated original oil in place, and ultimate recovery of crude oil in the United States, by type of entrapment. RESERVES CRUDE OIL, NATURAL GAS LIQUID AND NATURAL GAS

Crude oil, estimated proved reserves for the world and selected regions and countries, annual. BASIC PETROLEUM DATA BOOK

Crude oil, estimated reserves, by world region and country. INTERNATIONAL OIL DEVELOPMENTS

Crude oil, Federal Reserve Board indexes of industrial production, monthly and annual averages. STANDARD AND POOR'S TRADE AND SECURITIES STATIS-TICS

Crude oil, first recorded production by state. OIL PRODUCING INDUSTRY IN YOUR STATE

Crude oil, historical record of domestic production and proved reserves, na-tional totals as well as by state. RESERVES OF CRUDE OIL, NATURAL GAS LIQUID AND NATURAL GAS

Crude oil, producing wells drilled, domestic production, imports, consumption, exports, and year-end stocks, annual. STANDARD AND POOR'S TRADE AND SECURITIES STATISTICS

Crude oil, production and capacity in OAPEC AND OPEC countries, by country. INTERNATIONAL OIL DEVELOPMENTS

Crude oil, quantity passing through interstate pipeline transportation, annual. STANDARD AND POOR'S TRADE AND SECURITIES STATISTICS

Crude oil, refinery receipts, by method of transportation, daily averages. ANNUAL STATISTICAL REVIEW

Crude oil, refinery runs by selected world areas, annual. BASIC PETROLEUM DATA BOOK

Crude oil, refining capacity, in the world by country. INTERNATIONAL PETROLEUM

Crude oil, total transported in the United States, by method of transportation, annual. ENERGY STATISTICS

Crude oil, ultimate recovery and original oil in place by year of discovery, national totals as well as by state, annual. RESERVES OF CRUDE OIL, NATURAL GAS LIQUID AND NATURAL GAS

Crude oil, value, annual, historical data. OIL PRODUCING INDUSTRY IN YOUR STATE

Crude oil, value at wells, by state. OIL PRODUCING INDUSTRY IN YOUR STATE

Crude oil, value at wells produced at all times in each state. OIL PRODUC-ING INDUSTRY IN YOUR STATE

Crude oil, value produced, by state. OIL PRODUCING INDUSTRY IN YOUR STATE

Crude oil, year of largest production, by state. OIL PRODUCING INDUSTRY IN YOUR STATE

Crude oil and lease condensate, domestic, average value per barrel at the well, annual. ANNUAL STATISTICAL REVIEW

Crude oil and lease condensate, domestic production by state, daily averages. ANNUAL STATISTICAL REVIEW

Crude oil and lease condensate, domestic production, imports, new supply, disposition, exports, and use, daily averages. ANNUAL STATISTICAL REVIEW

Crude oil and lease condensate, domestic production and value at the well, annual. ANNUAL STATISTICAL REVIEW

Crude oil and lease condensate, domestic stocks by location, historical data, annual. ANNUAL STATISTICAL REVIEW

Crude oil and lease condensate, domestic supply and demand, daily averages. ANNUAL STATISTICAL REVIEW

Crude oil and petroleum products, barge movements, from PAD district III to PAD district II via the Mississippi River, daily averages. ANNUAL STATISTI-CAL REVIEW

Crude oil and petroleum products, tanker and barge movements, from U.S. Gulf Coast to the East Coast, daily averages. ANNUAL STATISTICAL RE-VIEW

Crude oil and petroleum products, tanker movements from the West Coast to the East Coast, daily averages. ANNUAL STATISTICAL REVIEW

Crude oil and refined products, conversion factors from metric tons to U.S. barrels, by country. INTERNATIONAL PETROLEUM

Crude oil and refined products, estimated imports of selected countries from Arab and non-Arab countries, by country. INTERNATIONAL OIL DEVELOP-MENTS

Crude oil and refined products, imports and exports, by country or area. BP STATISTICAL REVIEW OF THE WORLD OIL INDUSTRY

Crude oil pipelines, total mileage, annual. ENERGY STATISTICS

Crude oil price postings around the world by crude and gravity. BASIC PETRO-LEUM DATA BOOK

Crude oil prices, annual, historical data. OIL PRODUCING INDUSTRY IN YOUR STATE

Crude oil prices, monthly and annual averages. STANDARD AND POOR'S TRADE AND SECURITIES STATISTICS

Crude oil prices, worldwide. PLATT'S OIL PRICE HANDBOOK AND OIL-MANAC

Crude oil prices (posted). See also under specific companies, for example, Mobil Oil Corporation, crude oil prices (posted)

Crude oil prices (posted), and company cost, by area and type. STANDARD AND POOR'S INDUSTRY SURVEYS

Crude oil prices at oil fields in the United States, by state and company. PLATT'S OIL PRICE HANDBOOK AND OILMANAC

Crude oil prices in OPEC countries. INTERNATIONAL OIL DEVELOPMENTS

Crude oil prices in Venezuela, Canada, Persian Gulf, Libya, and Nigeria. INTERNATIONAL ECONOMIC REPORT OF THE PRESIDENT

Crude oil production. See also Crude OII, domestic production

Crude oil production, by country, daily averages. ANNUAL STATISTICAL RE-VIEW

Crude oil production, by country and region, in barrels per day. INTERNA-TIONAL OIL DEVELOPMENTS

Crude oil production, by select countries and regions, in barrels per day and as a percent of total production. INTERNATIONAL ECONOMIC REPORT OF THE PRESIDENT

Crude oil production, by world area and selected countries, annual. BASIC PETROLEUM DATA BOOK

Crude oil production, world total, by country or area. INTERNATIONAL PETROLEUM

Crude oil production for the free world, totals. BP STATISTICAL REVIEW OF THE WORLD OIL INDUSTRY

Crude oil production for the free world, totals by area. INTERNATIONAL PETROLEUM

Crude oil products, imports and exports of selected countries. ENERGY STATISTICS

Crude oil reserves. See also Crude oil domestic reserves

Crude oil reserves of the world by country. INTERNATIONAL PETROLEUM

Crude oil reserves of the world by country, estimated. ANNUAL STATISTICAL REVIEW

Crude oil resources, domestic and imported, of the United States, Japan, European Community, France, United Kingdom, West Germany, and Italy. INTERNATIONAL ECONOMIC REPORT OF THE PRESIDENT

Crude oil run by domestic refineries, by state, daily avaerages. ANNUAL STATISTICAL REVIEW

Crude petroleum, annual production in the United States. GAS FACTS

Crude petroleum, as a percent of total U.S. energy sources, annual. STANDARD AND POOR'S TRADE AND SECURITIES STATISTICS

Crude petroleum, average price at wells and by grade. TWENTIETH CENTURY PETROLEUM STATISTICS

Crude petroleum, average price at wells and by grade, by major exporting countries, and index of average prices. TWENTIETH CENTURY PETROLEUM STATISTICS

Crude petroleum, demand and supply, proved reserves, runs to stills, stocks, and value at wells. TWENTIETH CENTURY PETROLEUM STATISTICS

Crude petroleum, domestic production and runs to stills at refineries, monthly and annual. STANDARD AND POOR'S TRADE AND SECURITIES STATISTICS

Crude petroleum, domestic production by state, annual. STANDARD AND POOR'S INDUSTRY SURVEYS

Crude petroleum, exports and reexports by North America, Central America, and the Caribbean. INTERNATIONAL PETROLEUM

Crude petroleum, exports and reexports by the Eastern Hemisphere. INTERNATIONAL PETROLEUM

Crude petroleum, exports and reexports by the United States. INTERNATIONAL PETROLEUM

Crude petroleum, forecast of selected statistical data and annual growth rate for the United States. PREDICASTS

Crude petroleum, forecast of selected statistical data and annual growth rate for the world, by region and country. WORLD CASTS

Crude petroleum, imports by North America, Central America, and the Caribbean. INTERNATIONAL PETROLEUM

Crude petroleum, imports by South America. INTERNATIONAL PETROLEUM

Crude petroleum, imports by the Eastern Hemisphere. INTERNATIONAL PETRO-LEUM

Crude petroleum, imports by the Sino-Soviet area. INTERNATIONAL PETRO-LEUM

Crude petroleum, imports by the United States. INTERNATIONAL PETROLEUM

Crude petroleum, movement from exporters to importers, by country and region, annual. WORLD ENERGY SUPPLIES

Crude petroleum, movement throughout the world. INTERNATIONAL PETRO-LEUM

Crude petroleum, production, by world region and country, annual. WORLD ENERGY SUPPLIES

Crude petroleum, production, Eastern Hemisphere. INTERNATIONAL PETROLEUM

Crude petroleum, production, exports, imports, consumption by function and by user, and consumption for transformation in OECD member countries, annual. STATISTICS OF ENERGY

Crude petroleum, production, Western hemisphere. INTERNATIONAL PETRO-LEUM

Crude petroleum, production and consumption from water and nuclear power, annual. BITUMINOUS COAL DATA

Crude petroleum, production in the United States, annual. COAL FACTS; STANDARD AND POOR'S TRADE AND SECURITIES STATISTICS

Crude petroleum, production in the United States, by state. COAL FACTS

Crude petroleum, supply and demand by country or area. INTERNATIONAL PETROLEUM

Crude petroleum, wholesale price index, annual and monthly. PLATT'S OIL PRICE HANDBOOK AND OILMANAC

Crude petroleum, wholesale price per barrel, annual. NATIONAL PETROLEUM NEWS FACT BOOK

Crude petroleum and petroleum products carried in domestic transportation and percent of total arrived by each mode of transportation, daily averages. AN-NUAL STATISTICAL REVIEW

Crude runs to refineries in the United States, annual. STANDARD AND POOR'S INDUSTRY SURVEYS

Crude runs to stills by world region and selected countries. FINANCIAL ANALYSIS OF A GROUP OF PETROLEUM COMPANIES

Crude runs to stills in the United States, annual. STANDARD AND POOR'S TRADE AND SECURITIES STATISTICS

Cuba, crude production, percent of world production and reserves, refining capacity, tank ship fleet, deadweight tonnage, and tank ships under construction. TWENTIETH CENTURY PETROLEUM STATISTICS

Customers of gas and electric utilities in the United States. FINANCIAL STATISTICS OF PUBLIC UTILITIES

Czechoslovakia, crude production, percent of world reserves of crude and natural gas, refining capacity, and tank ship fleet. TWENTIETH CENTURY PETROLEUM STATISTICS

Czechoslovakia, oil and natural gas production, consumption and trade. INTERNATIONAL OIL DEVELOPMENTS

Daily average oil consumption, crude production, exports, imports, supply, refinery operations, and runs to stills. TWENTIETH CENTURY PETROLEUM STATISTICS

Dakotas, gasoline and fuel oil prices. OIL DAILY

Day's supply of stocks of oils, crude, and motor fuel. TWENTIETH CENTURY PETROLEUM STATISTICS

Dealer tank wagon gasoline prices in metropolitan areas. NATIONAL PETROLEUM NEWS FACT BOOK

Debenture issues of gas utilities by type of company, cost, and yield. GAS FACTS

Degree days, East Coast, by region and city, monthly. ANNUAL STATISTICAL REVIEW

Degree days, Great Lakes area, by region and city, monthly. ANNUAL STATISTICAL REVIEW

Degree days, Midwest area, by region and city, monthly. ANNUAL STATISTICAL REVIEW

Degree days, oil heating, monthly, by PAD district. MONTHLY ENERGY REVIEW

Degree days, southeast area by region and city, monthly. ANNUAL STATISTICAL REVIEW

Degree days, West Coast area and U.S. averages, by region and city, monthly. ANNUAL STATISTICAL REVIEW

Degree days for selected U.S. cities, thirty-year normals. GAS FACTS.

Delaware, installed generating capacity, electricity generation by type of prime mover, and electric energy sold by type of customer. EDISON ELECTRIC IN-STITUTE STATISTICAL YEARBOOK

Delaware, oil companies' market shares, by company. NATIONAL PETROLEUM NEWS FACT BOOK

Demand for energy fuels, by type, annual. TWENTIETH CENTURY PETROLEUM STATISTICS

Demand for energy input in the industrial, household, commercial, and transportation sectors in the United States, by type, annual. BASIC PETROLEUM DATA BOOK

Denmark, crude production, percent of world revenues of crude and natural gas, demand for refined products, refining capacity, tank ship fleet, deadweight tonnage, and tank ships under construction. TWENTIETH CENTURY PETROLEUM STATISTICS

Denmark, production, consumption, exports and imports by direction, of crude oil and crude oil products by type, and of natural gas and liquids. QUARTERLY OIL STATISTICS

Denmark, production, consumption, exports, and imports, by function and by user, of energy sources by type. STATISTICS OF ENERGY

Derby Refining Company, crude oil prices (posted). PLATT'S OIL PRICE HAND-BOOK AND OILMANAC

Development wells. See Oil wells; Wells

Diesel fuel prices in the United States and other OECD countries, in U.S. cents per gallon. INTERNATIONAL OIL DEVELOPMENTS

Diesel index gasoil, prices for Gulf Coast cargoes, high, low, and average. PLATT'S OIL PRICE HANDBOOK AND OILMANAC

Diesel (marine), refinery and terminal prices, high, low, and average. PLATT'S OIL PRICE HANDBOOK AND OILMANAC

Diesel oil sales in the United States by type, annual. BASIC PETROLEUM DATA BOOK

Discoveries. See also Wells, new; Oil wells, new; Gas wells, new

Discoveries of new oil and gas fields in the United States, total number and size, annual. BASIC PETROLEUM DATA BOOK

Distillate fuel oil, domestic consumption. MONTHLY ENERGY REVIEW

Distillate fuel oil, domestic demand, annual. NATIONAL PETROLEUM NEWS FACT BOOK

Distillate fuel oil, domestic demand, annual. STANDARD AND POOR'S IN-DUSTRY SURVEYS

Distillate fuel oil, domestic demand, export, import, percent refinery yield, and production, annual. TWENTIETH CENTURY PETROLEUM STATISTICS

Distillate fuel oil, domestic demand, production, and end-of-the-month stocks. STANDARD AND POOR'S TRADE AND SECURITIES STATISTICS

Distillate fuel oil, domestic demand, production, and stocks, annual. STAN-DARD AND POOR'S INDUSTRY SURVEYS

Distillate fuel oil, domestic demand and supply use, daily averages. ANNUAL STATISTICAL REVIEW

Distillate fuel oil, domestic production, annual. STANDARD AND POOR'S TRADE AND SECURITIES STATISTICS

Distillate fuel oil, domestic production, imports, and stocks. MONTHLY ENERGY REVIEW

Distillate fuel oil, forecast of selected statistical data and annual growth rates for the United States. PREDICASTS

Distillate fuel oil, forecast of selected statistical data and annual growth rates for the world, by region and country. WORLD CASTS

Distillate fuel oil, production, free-world percentages. INTERNATIONAL PETROLEUM

Distillate heating oil sales in the United States, by state and region. NA-TIONAL PETROLEUM NEWS FACT BOOK

Distillates, (petroleum) estimated consumption by state. OIL PRODUCING INDUSTRY IN YOUR STATE

Distillates, prices in selected cities, high, low, and average. PLATT'S OIL-MANAC

Distillates, spot prices in selected cities, high, low, and average. PLATT'S OIL PRICE HANDBOOK AND OILMANAC

Distillates and fuel oils, world cargo prices. PLATT'S OIL PRICE HANDBOOK AND OILMANAC

Distillates (light, middle), wholesale price index, annual and monthly. PLATT'S OIL PRICE HANDBOOK AND OILMANAC

Distillates (middle), inland demand for major products, by country or area. BP STATISTICAL REVIEW OF THE WORLD OIL INDUSTRY

District of Columbia, oil companies' market shares, by company. NATIONAL PETROLEUM NEWS FACT BOOK

Dividend rates, per share, utility stock, Moody's weighted averages, monthly and annual. MOODY'S PUBLIC UTILITY MANUAL

Dividends paid by gas and electric utilities in the United States. FINANCIAL STATISTICS OF PUBLIC UTILITIES; MOODY'S PUBLIC UTILITY MANUAL

Dominican Republic, demand for gasoline, kerosene, jet fuel, distillate, and residual fuel oil. TWENTIETH CENTURY PETROLEUM STATISTICS

Downstream service companies, estimated percentage sales to the petroleum industry. STANDARD AND POOR'S INDUSTRY SURVEYS

Dragline excavators of bituminous coal mines. BITUMINOUS COAL DATA

Drilling and equipping wells, by depth interval, estimated costs in the United States. ANNUAL STATISTICAL REVIEW

Drilling costs, average, in the United States, by type of well, annual. ANNUAL STATISTICAL REVIEW

Drilling costs, estimated, by well classification, onshore, in the United States, annual. BASIC PETROLEUM DATA BOOK

Drilling costs, footage drilled, new wells by state and PAD district, and wells drilled by type. TWENTIETH CENTURY PETROLEUM STATISTICS

Drilling operations and expenditures, by type of well. GAS FACTS

Dryers, automatic, gas, annual manufacturers' shipments. MOODY'S PUBLIC UTILITY MANUAL

Dry holes, average drilling costs in the United States, annual. ANNUAL STATISTICAL REVIEW

Dry holes, drilling and equipping, by depth interval, estimated costs in the United States. ANNUAL STATISTICAL REVIEW

Dry holes, expenditures by the domestic petroleum industry. CAPITAL INVESTMENTS IN THE PETROLEUM INDUSTRY

Dry holes, footage drilled and estimated drilling costs, by state. ANNUAL STATISTICAL REVIEW

Dry holes, number drilled, drilling cost, exploratory and wildcat wells. TWENTIETH CENTURY PETROLEUM STATISTICS

Dry holes, number of wells, footage drilled, and cost of drilling and equipping wells and dry holes, annual. ANNUAL STATISTICAL REVIEW

Dry holes, onshore, estimated drilling costs in the United States, annual. BASIC PETROLEUM DATA BOOK

Dry wells, by state. GAS FACTS

Dry wells, drilled in the United States, total number, annual. ANNUAL STATISTICAL REVIEW; STANDARD AND POOR'S TRADE AND SECURITIES STATISTICS

Dubai, crude oil production in barrels per day. INTERNATIONAL OIL DEVELOPMENTS

Dubai, proved reserves and percentage of world reserves of crude and natural gas. TWENTIETH CENTURY PETROLEUM STATISTICS

Duck bills and scrapes at underground bituminous coal mines. BITUMINOUS COAL DATA

Earnings, hourly and weekly, in mineral and other selected industries, annual. BITUMINOUS COAL DATA

Earnings, hourly average in crude petroleum, natural gas, and petroleum refining industry, annual. BASIC PETROLEUM DATA BOOK

Earnings per share of natural gas, transmission and distribution companies, Moody's weighted averages, quarterly. MOODY'S PUBLIC UTILITY MANUAL

Eastern Hemisphere, proved natural gas reserves, refining capacity, demand for refined products, tank ship fleet, deadweight tonnage, and tank ships under construction. TWENTIETH CENTURY PETROLEUM STATISTICS

East Germany, oil and natural gas production, consumption, and trade. INTERNATIONAL OIL DEVELOPMENTS

Ecuador, crude oil production in barrels per day. INTERNATIONAL OIL DEVELOPMENTS

Ecuador, crude oil production in barrels per day and as a percent of total production. INTERNATIONAL ECONOMIC REPORT OF THE PRESIDENT

Ecuador, crude production, percent of proved world reserves of crude and natural gas, demand for refined products, and refining capacity. TWENTIETH CENTURY PETROLEUM STATISTICS

Ecuador, imports from and exports to the United States and other developed countries, total value. INTERNATIONAL OIL DEVELOPMENTS

Ecuador, oil reserves and production, exports of crude oil and petroleum to the United States, annual. BASIC PETROLEUM DATA BOOK

Egypt, crude oil production, exports, percent of proved reserves of crude and natural gas, demand for refined products, refining capacity, tank ship fleet, deadweight tonnage, and tank ships under construction. TWENTIETH CENTURY PETROLEUM STATISTICS

Egypt, crude oil production in barrels per day. INTERNATIONAL OIL DEVELOPMENTS

Electrical energy loss distributed. MONTHLY ENERGY REVIEW

Electrical work in construction, by type of construction. STANDARD AND POOR'S INDUSTRY SURVEYS

Electric and gas utilities, total employment and payroll, annual. MOODY'S PUBLIC UTILITY MANUAL

Electric and gas utilities, weighted average yield on newly issued domestic bonds and preferred stocks, annual. EDISON ELECTRIC INSTITUTE STATISTICAL YEARBOOK

Electric energy production, by class of ownership, annual. STANDARD AND POOR'S INDUSTRY SURVEYS

Electric energy production, by source. BITUMINOUS COAL DATA

Electric energy sold to residential, commercial, industrial users, public authorities, and railroads, total, and investor-owned utilities, by state. EDISON ELECTRIC INSTITUTE STATISTICAL YEARBOOK

Electric generating plants, capacity and production by type of prime mover, annual. STANDARD AND POOR'S INDUSTRY SURVEYS

Electric generating plants, installed capacity of all types of prime movers by class of ownership, annual. MOODY'S PUBLIC UTILITY MANUAL

Electric generating plants, new or under construction, scheduled year of completion, and kilowatt capacity of new units. STEAM ELECTRIC PLANT FACTORS

Electric generating plants, thermal, hydro, and nuclear, installed capacity in industrial and public plants, by country, annual. WORLD ENERGY SUPPLIES

Electric generating plants, thermal, hydro, and nuclear, utilization of installed capacity in public and industrial plants, by country, annual. WORLD ENERGY SUPPLIES

Electric generation. See also Electric power production; Electricity generation; Electricity, production

Electric generation, analysis of fuels burned under boilers and by internal combustion engines, (coal, oil, and gas), average annual cost, by region. MOODY'S PUBLIC UTILITY MANUAL

Electricity. See also Electric power

Electricity, average residential retail prices, annual. GAS FACTS

Electricity, cost of, and comparison with the consumer price index. MOODY'S PUBLIC UTILITY MANUAL

Electricity, demand for input for electricity generation by utilities. BASIC PETROLEUM DATA BOOK

Electricity, domestic consumption through the year 2000 by consuming sector and as a percent of total energy input. BASIC PETROLEUM DATA BOOK

Electricity, domestic demand for input in industrial, household, commercial, and transportation sectors, annual. BASIC PETROLEUM DATA BOOK

Electricity, domestic production. See also Electricity, production; Electric power production; Electricity generation

Electricity, domestic production, by state. COAL FACTS

Electricity, domestic production by privately owned, municipal, federal, state, and cooperative utilities, annual. MOODY'S PUBLIC UTILITY MANUAL

Electricity, domestic sales, by month and class of service. EDISON ELECTRIC INSTITUTE STATISTICAL YEARBOOK

Electricity, domestic sales to customers by type, annual. EDISON ELECTRIC INSTITUTE STATISTICAL YEARBOOK

Electricity, domestic sales to public authorities, annual. EDISON ELECTRIC INSTITUTE STATISTICAL YEARBOOK

Electricity, domestic sales to residential, commercial, and industrial users, annual. MOODY'S PUBLIC UTILITY MANUAL

Electricity, domestic sales to small and large light and power, annual. EDISON ELECTRIC INSTITUTE STATISTICAL YEARBOOK

Electricity, domestic sales to ultimate consumer, annual. EDISON ELECTRIC INSTITUTE STATISTICAL YEARBOOK

Electricity, industrial and public, production, trade, and total and per capita consumption, by country, annual. WORLD ENERGY SUPPLIES

Electricity, net additions to generating capability, annual. STANDARD AND POOR'S INDUSTRY SURVEYS

Electricity, net kilowatts generated from coal, gas, and oil. STEAM ELECTRIC PLANT FACTORS

Electricity, production. See also Electricity, domestic production; Electricity generation; Electric generation

Electricity, production, imports, exports, and consumption, by function and user, in OECD member countries, annual. STATISTICS OF ENERGY

Electricity, production by industrial and public thermal, hydro, and nuclear plants, by country, annual. WORLD ENERGY SUPPLIES

Electricity, production from nuclear and water power. TWENTIETH CENTURY PETROLEUM STATISTICS

Electricity, production in selected countries, annual. INTERNATIONAL ECONOMIC REPORT OF THE PRESIDENT

Electricity, residential users, total number, average use, average annual bill, average revenue per kilowatt hour, and annual growth rates. STANDARD AND POOR'S INDUSTRY SURVEYS

Electricity bills. See Electric power bills

Electricity customers (ultimate), by class of service and average number of customers, by state. EDISON ELECTRIC INSTITUTE STATISTICAL YEARBOOK

Electricity from hydro and nuclear power, domestic consumption and production, annual. COAL FACTS; GAS FACTS

Electricity from hydro and nuclear power, world production by region and country, annual. WORLD ENERGY SUPPLIES

Electricity from water power, domestic consumption and production, annual. COAL FACTS

Electricity generated by average BTU value of coal, oil, and gas. STEAM ELECTRIC PLANT FACTORS

Electricity generation. See also Electricity, production

Electricity generation by type of prime mover, by kind of fuel and type of ownership, by state. EDISON ELECTRIC INSTITUTE STATISTICAL YEARBOOK

Electricity generation in the United States by type of ownership, annual. EDISON ELECTRIC INSTITUTE STATISTICAL YEARBOOK

Electricity generation (net), by individual electric utilities. STEAM ELECTRIC PLANT FACTORS

Electricity lost and unaccounted for, between generation and ultimate disposition, annual. EDISON ELECTRIC INSTITUTE STATISTICAL YEARBOOK

Electric light and power construction, cost trends by region. EDISON ELECTRIC INSTITUTE STATISTICAL YEARBOOK

Electric light and power industry, capacity, peak load, production, average kilowatt hour per customer, and average revenue per kilowatt hour, annual. STANDARD AND POOR'S TRADE AND SECURITIES STATISTICS

Electric light and power industry, generator capacity, kilowatt hours generated, fuel used, sales revenues, number of customers by type, and average annual use and bill per residential customer. MOODY'S PUBLIC UTILITY MANUAL

Electric light and power utilities, new construction expenditures, annual. STANDARD AND POOR'S INDUSTRY SURVEYS

Electric light and power utilities, new constructions, cost trends, by region, annual. MOODY'S PUBLIC UTILITY MANUAL

Electric lighting and power customers central stations by type of service, annual. MOODY'S PUBLIC UTILITY MANUAL

Electric line (overhead), circuit miles and voltage, by state. EDISON ELECTRIC INSTITUTE STATISTICAL YEARBOOK

Electric operating companies, independent, comparative income and balance sheet data on selected companies by size. MOODY'S PUBLIC UTILITY MANUAL

Electric operating expenses of investor-owned utilities, annual. EDISON ELECTRIC INSTITUTE STATISTICAL YEARBOOK

Electric output, weekly index of all electric utility industry, annual. EDISON ELECTRIC INSTITUTE STATISTICAL YEARBOOK

Electric output, weekly totals of all electric utility industry. EDISON ELECTRIC INSTITUTE STATISTICAL YEARBOOK

Electric power. See also Electricity

Electric power, average kilowatt hour used per customer by type and year. EDISON ELECTRIC INSTITUTE STATISTICAL YEARBOOK

Electric power, capability and peak load, annual. EDISON ELECTRIC INSTITUTE STATISTICAL YEARBOOK

Electric power, consumption for energy generation, annual. BASIC PETROLEUM DATA BOOK

Electric power, estimated net generation, and cost per kilowatt hour, by type of fuel consumed, by region, annual. MOODY'S PUBLIC UTILITY MANUAL

Electric power, exports to Canada and Mexico, annual. EDISON ELECTRIC INSTITUTE STATISTICAL YEARBOOK

Electric power, imports from Canada and Mexico, annual. EDISON ELECTRIC INSTITUTE STATISTICAL YEARBOOK

Electric power, kilowatt hour per person per year. EDISON ELECTRIC INSTITUTE STATISTICAL YEARBOOK

Electric power, output by all electric utilities, annual. EDISON ELECTRIC INSTITUTE STATISTICAL YEARBOOK

Electric power, sales by type of customer, annual. STANDARD AND POOR'S INDUSTRY SURVEYS

Electric power, sales to ultimate consumers, annual. EDISON ELECTRIC INSTITUTE STATISTICAL YEARBOOK

Electric power, source and disposition in the United States, annual. EDISON ELECTRIC INSTITUTE STATISTICAL YEARBOOK

Electric power bills, average annual amount by type of customer. EDISON ELECTRIC INSTITUTE STATISTICAL YEARBOOK

Electric power bills, residential, weighted annual average and total number of customers. MOODY'S PUBLIC UTILITY MANUAL

Electric power production. See also Electricity generation; Electricity, production

Electric power production, monthly and mean daily output. STANDARD AND POOR'S TRADE AND SECURITIES STATISTICS

Electric utilities, allocation of utility revenues, annual. STANDARD AND POOR'S INDUSTRY SURVEYS

Electric utilities, average revenues per kilowatt hour sold, by class of customers. EDISON ELECTRIC STATISTICAL YEARBOOK

Electric utilities, capability, peak load and reserve margins, annual. STANDARD AND POOR'S INDUSTRY SURVEYS

Electric utilities, capability additions, new, by type of power, annual. MOODY'S PUBLIC UTILITY MANUAL

Electric utilities, composite financial and related statistics, annual. STANDARD AND POOR'S INDUSTRY SURVEYS

Electric utilities, composite income statement, annual. MOODY'S PUBLIC UTILITY MANUAL

Electric utilities, composite income statement, annual. MOODY'S PUBLIC UTILITY MANUAL

Electric utilities, composite operating revenue and expenses, income, taxes, depreciation, long-term debt, and dividends, annual. MOODY'S PUBLIC UTILITY MANUAL

Electric utilities, construction expenditures by type, annual. STANDARD AND POOR'S INDUSTRY SURVEYS

Electric utilities, construction expenditures on steam, hydro, and pumped storage, transmission and distribution plants, annual. MOODY'S PUBLIC UTILITY MANUAL

Electric utilities, consumption of fossil fuel for electric generation, by type of fuel. STANDARD AND POOR'S INDUSTRY SURVEYS

Electric utilities, directory, showing name of the company and the city served. MOODY'S PUBLIC UTILITY MANUAL

Electric utilities, domestic, income account, balance sheet, financial and operating ratios, capitalization, area served, customers, and kilowatts sold. FINANCIAL STATISTICS OF PUBLIC UTILITIES

Electric utilities, electricity generation per kilowatt of installed name plate capacity, by ownership and type of prime mover, annual. EDISON ELECTRIC INSTITUTE STATISTICAL YEARBOOK

Electric utilities, end-of-the-month coal and oil stocks. MONTHLY ENERGY REVIEW

Electric utilities, Federal Reserve Board indexes of industrial production, monthly and annual averages. STANDARD AND POOR'S TRADE AND SECURITIES STATISTICS

Electric utilities, forecast of selected statistical data and annual growth rates for the United States. PREDICASTS

Electric utilities, forecast of selected statistical data and annual growth rates for the world, by country and region. WORLD CASTS

Electric utilities, fuel consumption for electric generation. MONTHLY ENERGY REVIEW

Electric utilities, installed capacity of generating plants of privately owned

municipal, federal, state, and cooperative utilities, annual. MOODY'S PUB-
LIC UTILITY MANUAL

Electric utilities, installed generating capacity by type of prime mover, by
state. EDISON ELECTRIC INSTITUTE STATISTICAL YEARBOOK

Electric utilities, investor-owned. See also Electric utilities, privately owned

Electric utilities, investor-owned, detailed composite income statement and
operating expenses, annual. MOODY'S PUBLIC UTILITY MANUAL

Electric utilities, investor-owned, security sales, bonds, common and prepared
stocks, annual. MOODY'S PUBLIC UTILITY MANUAL

Electric utilities, mode of transportation of coal by individual utilities, by
state. COAL TRAFFIC ANNUAL

Electric utilities, privately owned. See also Electric utilities, investor-owned

Electric utilities, privately owned, balance sheet and income account items for
individual utilities. STATISTICS OF PRIVATELY OWNED UTILITIES

Electric utilities, privately owned, composite assets and liabilities, annual.
MOODY'S PUBLIC UTILITY MANUAL

Electric utilities, privately owned, composite income account and balance sheet.
STATISTICS OF PRIVATELY OWNED UTILITIES

Electric utilities, privately owned, customers by type, annual. STATISTICS OF
PRIVATELY OWNED UTILITIES

Electric utilities, privately owned, generating stations, by type of prime mover,
and generating capacity for individual utilities. STATISTICS OF PRIVATELY
OWNED UTILITIES

Electric utilities, privately owned, generating stations by type of prime mover,
and source and disposition of electrical energy. STATISTICS OF PRIVATELY
OWNED UTILITIES

Electric utilities, privately owned, number of customers, kilowatt hour sales,
and revenue for individual utilities. STATISTICS OF PRIVATELY OWNED
UTILITIES

Electric utilities, privately owned, operating revenues by type. STATISTICS
OF PRIVATELY OWNED UTILITIES

Electric utilities, privately owned, operation and maintenance expenses.
STATISTICS OF PRIVATELY OWNED UTILITIES

Electric utilities, privately owned, power production expenses, by type of
power. STATISTICS OF PRIVATELY OWNED UTILITIES

Electric utilities, privately owned, sales of firm electric power for resale by regions. STATISTICS OF PUBLICLY OWNED UTILITIES

Electric utilities, privately owned, selected composite balance sheet items, annual. MOODY'S PUBLIC UTILITY MANUAL

Electric utilities, privately owned, transmission, distribution, sales and administrative expenses. STATISTICS OF PRIVATELY OWNED UTILITIES

Electric utilities, privately owned, utility plant operation and maintenance expenses and electric energy account for individual utilities. STATISTICS OF PRIVATELY OWNED UTILITIES

Electric utilities, production (net) through coal, oil, gas, nuclear, hydroelectric, and other sources. MONTHLY ENERGY REVIEW

Electric utilities, publicly owned, balance sheet and income account items for individual utilities. STATISTICS OF PUBLICLY OWNED UTILITIES

Electric utilities, publicly owned, composite income account and balance sheet. STATISTICS OF PUBLICLY OWNED UTILITIES

Electric utilities, publicly owned, generating stations by type of prime mover, and generating capacity for individual utilities. STATISTICS OF PUBLICLY OWNED UTILITIES

Electric utilities, publicly owned, generating stations by type of prime mover, and source and disposition of electrical energy. STATISTICS OF PUBLICLY OWNED UTILITIES

Electric utilities, publicly owned, kilowatt hour sales by type of customers. STATISTICS OF PUBLICLY OWNED UTILITIES

Electric utilities, publicly owned, number of customers, kilowatt hour sales, and revenue for individual utilities. STATISTICS OF PUBLICLY OWNED UTILITIES

Electric utilities, publicly owned, operating revenues by type. STATISTICS OF PUBLICLY OWNED UTILITIES

Electric utilities, publicly owned, operation and maintenance expenses. STATISTICS OF PUBLICLY OWNED UTILITIES

Electric utilities, publicly owned, power production expenses, by type of power. STATISTICS OF PUBLICLY OWNED UTILITIES

Electric utilities, publicly owned, sales of firm electric power for resale by regions. STATISTICS OF PUBLICLY OWNED UTILITIES

Electric utilities, publicly owned, total customers by type, annual. STATISTICS OF PUBLICLY OWNED UTILITIES

Electric utilities, publicly owned, transmission, distribution, sales, and administrative expenses. STATISTICS OF PUBLICLY OWNED UTILITIES

Electric utilities, publicly owned, utility plant operation and maintenance expenses and electric energy account for individual utilities. STATISTICS OF PUBLICLY OWNED UTILITIES

Electric utilities, sales to residential, commercial, industrial, and other customers. MONTHLY ENERGY REVIEW

Electric utilities, selected financial ratios and relationships, annual. MOODY'S PUBLIC UTILITY MANUAL

Electric utilities stocks, capitalization in terms of long- and short-term debt and common stocks, annual. MOODY'S PUBLIC UTILITY MANUAL

Electric utilities stocks, composite book value per share, annual. MOODY'S PUBLIC UTILITY MANUAL

Electric utilities stocks, composite dividend per share, annual. MOODY'S PUBLIC UTILITY MANUAL

Electric utilities stocks, composite earnings per share, annual. MOODY'S PUBLIC UTILITY MANUAL

Electric utilities stocks, composite pay-out ratio per share, annual. MOODY'S PUBLIC UTILITY MANUAL

Electric utilities stocks, return on equity, composite rate, annual. MOODY'S PUBLIC UTILITY MANUAL

Electric utility industry, installed generating capacity, by ownership and type of prime mover and by state. EDISON ELECTRIC INSTITUTE STATISTICAL YEARBOOK

Electric utility industry, revenue and growth rates, by type of user, annual. STANDARD AND POOR'S INDUSTRY SURVEYS

Electric utility industry, revenues by class of service, by month. EDISON ELECTRIC INSTITUTE STATISTICAL YEARBOOK

Electric utility industry, sales by type of customer, monthly and annual. STANDARD AND POOR'S INDUSTRY SURVEYS

Electric utility industry, ultimate customers by year and class of service and by state. EDISON ELECTRIC INSTITUTE STATISTICAL YEARBOOK

Employment, earnings, and payrolls in petroleum manufacturing industries in the United States. ANNUAL STATISTICAL REVIEW

Employment and earnings in the gas utility industry, annual. MOODY'S PUBLIC UTILITY MANUAL

Employment at bituminous coal mines by type of mine. BITUMINOUS COAL DATA

Employment at coal mines by state and district. BITUMINOUS COAL DATA

Employment in crude oil and natural gas production, by state. OIL PRODUCING INDUSTRY IN YOUR STATE

Employment in gas companies and systems, by state. OIL PRODUCING INDUSTRY IN YOUR STATE

Employment in gasoline service stations, by state. OIL PRODUCING INDUSTRY IN YOUR STATE

Employment in operation, maintenance, and construction in investor-owned utilities, annual. EDISON ELECTRIC INSTITUTE STATISTICAL YEARBOOK

Employment in petroleum and natural gas extraction and petroleum refining industries, annual. BASIC PETROLEUM DATA BOOK

Employment in petroleum industry by state. OIL PRODUCING INDUSTRY IN YOUR STATE

Employment in petroleum refining by state. OIL PRODUCING INDUSTRY IN YOUR STATE

Employment in pipeline transportion by state. OIL PRODUCING INDUSTRY IN YOUR STATE

Energy. See also under specific fuels

Energy, commercial, aggregate, per capita, world, regional, and by country consumption, annual. WORLD ENERGY SUPPLIES

Energy, commercial, world and regional, and by country imports and exports, annual. WORLD ENERGY SUPPLIES

Energy, domestic consumption by major resource and by consuming sector, annual. GAS FACTS

Energy, domestic consumption, by major resource and by consuming sector, annual. ENERGY STATISTICS

Energy, domestic consumption, total consumed. BP STATISTICAL REVIEW OF THE WORLD OIL INDUSTRY

Energy, domestic consumption by sector and by primary source. MONTHLY ENERGY REVIEW

Energy, domestic consumption by source and by consuming sector, through the year 2000. BASIC PETROLEUM DATA BOOK

Energy, domestic consumption per capita, annual. GAS FACTS

Energy, domestic production by major source, annual. BASIC PETROLEUM DATA BOOK

Energy, domestic supply by type, estimates. STANDARD AND POOR'S INDUS-TRY SURVEYS

Energy, forecast of selected statistical data and annual growth rates for the United States. PREDICASTS

Energy, forecast of selected statistical data and annual growth rates for the world, by region and country. WORLD CASTS

Energy, imports (net) in Japan. BP STATISTICAL REVIEW OF THE WORLD OIL INDUSTRY

Energy, imports (net) in the United States. BP STATISTICAL REVIEW OF THE WORLD OIL INDUSTRY

Energy, imports (net) in Western Europe. BP STATISTICAL REVIEW OF THE WORLD OIL INDUSTRY

Energy, volume utilized for generation by type of energy, annual. BASIC PETROLEUM DATA BOOK

Energy consumption, by world region, through 1990, and by source. BASIC PETROLEUM DATA BOOK

Energy consumption, primary, by the world, by country or area. BP STATISTI-CAL REVIEW OF THE WORLD OIL INDUSTRY

Energy consumption, totals in Western Europe. BP STATISTICAL REVIEW OF THE WORLD OIL INDUSTRY

Energy fuels, domestic consumption, annual. COAL FACTS

Energy input, gross and net, in the domestic economy, annual. BASIC PETROLEUM DATA BOOK

Energy per capita, gross and net in the United States, annual. BASIC PETROLEUM DATA BOOK

Energy resources, by major source and consuming sector, domestic gross con-sumption, daily averages. ANNUAL STATISTICAL REVIEW

Energy resources, domestic, percentages by type of source, annual. STANDARD AND POOR'S TRADE AND SECURITIES STATISTICS

Energy resources, domestic and imported by the United States, European community, Japan, United Kingdom, France, West Germany, and Italy. INTERNATIONAL ECONOMIC REPORT OF THE PRESIDENT

Ethane, forecast of selected statistical data and annual growth rates for the United States. PREDICASTS

Ethane, forecast of selected statistical data and annual growth rates for the world, by region and country. WORLD CASTS

Europe, crude production, percent of world proved crude and natural gas reserves, demand for refined products, refining capacity, tank ship fleet, deadweight tonnage, and tank ships under construction. TWENTIETH CENTURY PETROLEUM STATISTICS

Europe, production and use of energy sources by type of source and by country. STATISTICS OF ENERGY

European Economic Community, primary energy production, annual. INTERNATIONAL ECONOMIC REPORT OF THE PRESIDENT

European Economic Community, production, consumption, imports and exports by direction, of crude oil and crude oil products by type, and of natural gas and liquids. QUARTERLY OIL STATISTICS

European Economic Community, production and use of energy sources by type of energy and by country. STATISTICS OF ENERGY

Exploration, geophysical, in the United States and selected countries, annual. BASIC PETROLEUM DATA BOOK

Exploration and development expenditures, by the domestic petroleum industry. CAPITAL INVESTMENTS IN THE PETROLEUM INDUSTRY; FINANCIAL ANALYSIS OF A GROUP OF PETROLEUM COMPANIES.

Exploratory activity in the United States, monthly, crews engaged and line miles of seismic exploration. MONTHLY ENERGY REVIEW

Exploratory wells. See Wells; Oil wells

Exports. See also under specific energy sources

Exports of all oils, crude, fuel oil, kerosene, motor fuel, and refined products. TWENTIETH CENTURY PETROLEUM STATISTICS

Exports of coal and anthracite by the United States to other countries, by country, region, volume, and value. WORLD COAL TRADE

Exports of electricity to Canada and Mexico, annual. EDISON ELECTRIC INSTITUTE STATISTICAL YEARBOOK

Exports to the OPEC countries from developed countries, by country, total value. INTERNATIONAL OIL DEVELOPMENTS

Exxon Company (U.S.), crude oil prices (posted). PLATT'S OIL PRICE HANDBOOK AND OILMANAC

Far East, capital and exploration expenditures and gross and net investments in fixed assets, by the domestic petroleum industry. CAPITAL INVESTMENTS IN THE PETROLEUM INDUSTRY

Far East, export of solid fuel to other countries, by country. WORLD ENERGY SUPPLIES

Far Eastern exports of crude petroleum to other countries, by country, annual. WORLD ENERGY SUPPLIES

Fast breeder reactors, description, and net electrical power generation, by country. POWER REACTORS IN MEMBER STATES

Federal Reserve Board indexes of industrial production for the energy sector by sector, monthly and annual averages. STANDARD AND POOR'S TRADE AND SECURITIES STATISTICS

Federal utilities. See Electric utilities, publicly owned

Feedstocks, petrochemical, forecast of selected statistical data and annual growth rates for the United States. PREDICASTS

Feedstocks, petrochemical, forecast of selected statistical data and annual growth rates for the world, by region and country. WORLD CASTS

Feedstocks, production, imports, exports, consumption, and stocks in OECD countries, annual and quarterly. QUARTERLY OIL STATISTICS

Finland, production, consumption, exports and imports by direction, of crude oil and crude oil products by type, and of natural gas and liquids. QUARTERLY OIL STATISTICS

Finland, production, consumption, exports, and imports, by function and user, of energy sources by type. STATISTICS OF ENERGY

Floor and direct heating equipment, annual manufacturers' shipments. MOODY'S PUBLIC UTILITY MANUAL

Floor and wall furnaces, annual manufacturers' shipments. GAS FACTS

Florida, crude production and reserves, daily average production per well, footage drilled, natural gas reserves, natural gasoline production at plants, well completions, and producing oil wells. TWENTIETH CENTURY PETROLEUM STATISTICS

Florida, installed generating capacity, electricity generation by type of prime mover, and electric energy sold by type of customer. EDISON ELECTRIC INSTITUTE STATISTICAL YEARBOOK

Florida, oil and natural gas production, reserves, exploration, development, prices, and related data, annual. OIL PRODUCING INDUSTRY IN YOUR STATE

Florida, oil companies' market shares, by company. NATIONAL PETROLEUM NEWS FACT BOOK

Florida, total oil and gas wells, dry holes, development and exploratory wells drilled, by quarter. QUARTERLY REVIEW OF DRILLING STATISTICS

Forecasts. See under specific items, for example, Coal, forecast of selected statistical data

Fossil fuels, consumption for electricity generation annually and by kind of fuel. EDISON ELECTRIC INSTITUTE STATISTICAL YEARBOOK

Fossil steam-electric utilities, reported new capability addition, by type of ownership. MOODY'S PUBLIC UTILITY MANUAL

France, crude production, percent of world proved reserves of crude and natural gas, demand for refined products, refining capacity, tank ship fleet; deadweight tonnage, and tank ships under construction. TWENTIETH CENTURY PETROLEUM STATISTICS

France, estimated imports of crude oil and refined products from Arab and non-Arab countries by original crude source. INTERNATIONAL OIL DEVELOPMENTS

France, imports and exports of crude and crude products. INTERNATIONAL OIL DEVELOPMENTS

France, primary energy production, annual. INTERNATIONAL ECONOMIC REPORT OF THE PRESIDENT

France, production, consumption, exports and imports by direction, of crude oil and crude oil products by type, and of natural gas and liquids. QUARTERLY OIL STATISTICS

France, production, consumption, exports, and imports, by type. STATISTICS OF ENERGY

Freight and spot rates from the Middle East to northwest Europe. BP STATISTICAL REVIEW OF THE WORLD OIL INDUSTRY

Fuel and lighting, wholesale price indexes. BITUMINOUS COAL DATA

Fuel consumption by electric utilities and unit costs, a comparative summary by region. STEAM ELECTRIC PLANT FACTORS

Fuel consumption by electric utilities for electricity generation and unit costs. STEAM ELECTRIC PLANT FACTORS

Fuel consumption by mode of transportation, annual. ENERGY STATISTICS

Fuel consumption by passenger cars, annual averages. NATIONAL PETROLEUM NEWS FACT BOOK

Fuel consumption by type of vehicle and average miles per gallon. NATION-AL PETROLEUM NEWS FACT BOOK

Fuel consumption of utilities. ANNUAL STATISTICAL REVIEW

Fuel cost of consumption of oil, coal, and gas for electric generation under broilers and internal combustion engines, by state. EDISON ELECTRIC INSTITUTE STATISTICAL YEARBOOK

Fuel cost to electric utilities, by type of fuel, annual. STANDARD AND POOR'S INDUSTRY SURVEYS

Fuel-generated electricity, amount per kind of fuel, by state, annual. EDI-SON ELECTRIC INSTITUTE STATISTICAL YEARBOOK

Fuel (light and heavy), average wholesale prices, historical data. PLATT'S OIL PRICE HANDBOOK AND OILMANAC

Fuel (light and heavy), wholesale prices per gallon, annual. NATIONAL PETROLEUM NEWS FACT BOOK

Fuel oil. See also Distillate fuel oil; Residual fuel oil

Fuel oil, average annual residential retail prices. GAS FACTS

Fuel oil, distillate and residual, contribution to freight revenues of U.S. railroads, annual. ENERGY STATISTICS

Fuel oil, distillate and residual, domestic demand, annual. TWENTIETH CENTURY PETROLEUM STATISTICS

Fuel oil, distillate and residual, domestic demand and supply. BASIC PETRO-LEUM DATA BOOK

Fuel oil, distillate and residual, domestic demand by type of use. NA-TIONAL PETROLEUM NEWS FACT BOOK

Fuel oil, distillate and residual, domestic imports, total and by PAD district, annual. BASIC PETROLEUM DATA BOOK

Fuel oil, distillate and residual, domestic sales by type of customer. BITUMI-NOUS COAL DATA

Fuel oil, distillate and residual, production, trade, total and per capita apparent consumption, by country, annual. WORLD ENERGY SUPPLIES

Fuel oil, distillate and residual, sales by state and by type of use. BASIC PETROLEUM DATA BOOK

Fuel oil, distillate and residual, sales to gas and electric public utility power plants. BASIC PETROLEUM DATA BOOK

Fuel oil, domestic consumption, annual. STANDARD AND POOR'S TRADE AND SECURITIES STATISTICS

Fuel oil, domestic demand, annual, for all oils. STANDARD AND POOR'S INDUSTRY SURVEYS

Fuel oil, domestic production, demand, and end-of-the-month stocks. STANDARD AND POOR'S TRADE AND SECURITIES STATISTICS

Fuel oil, domestic production, demand, and stocks, annual. STANDARD AND POOR'S INDUSTRY SURVEYS

Fuel oil, domestic sales by state and region. NATIONAL PETROLEUM NEWS FACT BOOK

Fuel oil, inland demand for major products, by country or area. BP STATISTICAL REVIEW OF THE WORLD OIL INDUSTRY

Fuel oil consumption for electricity generation, annual. EDISON ELECTRIC INSTITUTE STATISTICAL YEARBOOK

Fuel oil dealers, total number and sales by state and region. NATIONAL PETROLEUM NEWS FACT BOOK

Fuel-oil-generated electricity, total amount by state, annual. EDISON ELECTRIC INSTITUTE STATISTICAL YEARBOOK

Fuel oil prices, monthly and annual averages, at New York Harbor. STANDARD AND POOR'S TRADE AND SECURITIES STATISTICS

Fuel oil prices at New York by type of fuel oil, annual. STANDARD AND POOR'S INDUSTRY SURVEYS

Fuel oil prices in selected cities and states. OIL DAILY

Fuel oil (residual) production, export, import, consumption, and supply in OECD countries, annual and quarterly. QUARTERLY OIL STATISTICS

Fuel oil tax revenues of state and federal governments. NATIONAL PETROLEUM NEWS FACT BOOK

Fuel (patent), production, imports, exports, consumption by function and user, and consumption for transformation in OECD member countries, annual. STATISTICS OF ENERGY

Fuel rate at steam-electric plants for selected years, by region. STEAM ELECTRIC PLANT FACTORS

Fuels, solid and liquid, world, regional, and by country, consumption, annual. WORLD ENERGY SUPPLIES

Fuels (mineral energy), demand and consumption in the United States, in BTUs. TWENTIETH CENTURY PETROLEUM STATISTICS

Fuel used for clothes dryers among urban and rural occupied units. GAS FACTS

Fuel used for cooking, distribution among urban occupied units, by state. GAS FACTS

Fuel utilization by steam-electric plants, trends in efficiency. STEAM ELECTRIC PLANT FACTORS

Furnaces, wall, annual manufacturers' shipments. MOODY'S PUBLIC UTILITY MANUAL

Furnaces, warm air, annual manufacturers' shipments. GAS FACTS; MOODY'S PUBLIC UTILITY MANUAL

Gabon, crude oil production, percent of proved world reserves of crude oil and natural gas, and demand for refined products. TWENTIETH CENTURY PETROLEUM STATISTICS

Gabon, crude oil production in barrels per day. INTERNATIONAL OIL DEVELOPMENTS

Gabon, crude oil production in barrels per day and as a percent of total production. INTERNATIONAL ECONOMIC REPORT OF THE PRESIDENT

Gabon, imports from and exports to the United States and other developed countries, total value. INTERNATIONAL OIL DEVELOPMENTS

Gabon, oil reserves and production, exports of crude oil and petroleum to the United States, annual. BASIC PETROLEUM DATA BOOK

Gas, indexes of wholesale prices, yearly and monthly averages. GAS FACTS

Gas, production, imports, exports, consumption by function and user, and consumption for transformation in OECD member countries, annual. STATISTICS OF ENERGY

Gas, underground storage fields, number of pools, wells, compressor stations, horsepower and ultimate capacity, by state. GAS FACTS

Gas and condensate wells. See also Gas wells; Wells, gas

Gas and condensate wells in the United States, producing wells, total number, by state, annual. BASIC PETROLEUM DATA BOOK

Gas appliances, residential, manufacturers' shipments, annual. GAS FACTS

Gas as a percentage of total U.S. energy resources, annual. STANDARD AND POOR'S TRADE AND SECURITIES STATISTICS

Gas (blast furnace), production, imports, exports, consumption by function and user, and consumption for transformation in OECD member countries, annual. STATISTICS OF ENERGY

Gas (coal equivalent) consumption by steam-electric plant, by region. COAL FACTS

Gas coke, production, imports, exports, consumption by function and user, and consumption for transformation in OECD member countries, annual. STATISTICS OF ENERGY

Gas consumption by residential appliances, by region. GAS FACTS

Gas consumption for electricity generation, by individual utility, and unit costs. STEAM ELECTRIC PLANT FACTORS

Gas-cooled graphite-moderated reactors, description and net electrical power generation, by country. POWER REACTORS IN MEMBER STATES

Gas/diesel conversion factors. BP STATISTICAL REVIEW OF THE WORLD OIL INDUSTRY

Gas dryers, automatic, manufacturers' shipments, annual. GAS FACTS

Gas-generated electricity, total amount, annual and by state. EDISON ELEC-TRIC INSTITUTE STATISTICAL YEARBOOK

Gas-heated housing units, total number, by census division. MOODY'S PUB-LIC UTILITY MANUAL

Gas hearing equipment, manufacturers' shipments, annual. GAS FACTS

Gas heating equipment, manufacturers' shipments by class and type of equip-ment. NATIONAL PETROLEUM NEWS FACT BOOK

Gas heating equipment, manufacturers' shipments by type, annual. MOODY'S PUBLIC UTILITY MANUAL

Gas (native) in storage reservoirs, total volume by state. GAS FACTS

Gas in underground storage, inputs by state. GAS FACTS

Gas in underground storage, outputs by state. GAS FACTS

Gasoil, European bulk prices. PLATT'S OIL PRICE HANDBOOK AND OIL-MANAC

Gas oil, production, domestic demand, and end-of-the-month stocks. STANDARD AND POOR'S TRADE AND SECURITIES STATISTICS

Gas oil, production, exports, imports, and total internal consumption in OECD member countries, annual. STATISTICS OF ENERGY

Gas oil, production, exports, imports, consumption, and supply in OECD member countries, annual and quarterly. QUARTERLY OIL STATISTICS

Gasoline. See also Motor gasoline

Gasoline, contribution to freight revenue of U.S. railroad companies, annual. ENERGY STATISTICS

Gasoline, domestic consumption, annual. STANDARD AND POOR'S TRADE AND SECURITIES STATISTICS

Gasoline, domestic consumption by state and ranking. NATIONAL PETROLEUM NEWS FACT BOOK

Gasoline, domestic demand, annual. NATIONAL PETROLEUM NEWS FACT BOOK; STANDARD AND POOR'S INDUSTRY SURVEYS

Gasoline, domestic demand, production, and percent refinery yield. TWENTIETH CENTURY PETROLEUM STATISTICS

Gasoline, domestic demand and supply, daily averages, annual. BASIC PETROLEUM DATA BOOK

Gasoline, domestic production. See also Gas production; Gasoline production

Gasoline, domestic production, annual. STANDARD AND POOR'S TRADE AND SECURITIES STATISTICS

Gasoline, domestic production, exports, and end-of-the-month stocks, monthly. STANDARD AND POOR'S TRADE AND SECURITIES STATISTICS

Gasoline, domestic production, imports, new supply, stocks, consumption, and exports, annual. STANDARD AND POOR'S TRADE AND SECURITIES STATISTICS

Gasoline, domestic production, stocks, demand, consumption, and exports, annual. STANDARD AND POOR'S INDUSTRY SURVEYS

Gasoline, domestic supply and demand by use, daily average. ANNUAL STATISTICAL REVIEW

Gasoline, European bulk prices. PLATT'S OIL PRICE HANDBOOK AND OIL-MANAC

Gasoline, forecast of selected statistical data and annual growth rates for the United States. PREDICASTS

Gasoline, forecast of selected statistical data and annual growth rates for the world, by region and country. WORLD CASTS

Gasoline, inland demand for major products, by country or area. BP STATISTI-CAL REVIEW OF THE WORLD OIL INDUSTRY

Gasoline, market sales by company, nationally and by state. NATIONAL PETROLEUM NEWS FACT BOOK

Gasoline, motor and aviation, domestic demand, annual. ENERGY STATISTICS

Gasoline, national, world, and regional production, supply, imports, and exports, annual. WORLD ENERGY SUPPLIES

Gasoline, production, stocks, imports, exports, and domestic demand in the United States, annual. ENERGY STATISTICS

Gasoline (aviation, motor, and natural), production, trade, and total and per capita apparent consumption, by country, annual. WORLD ENERGY SUPPLIES

Gasoline brand names, by company. NATIONAL PETROLEUM NEWS FACT BOOK

Gasoline brands marketed in all states by company. NATIONAL PETROLEUM NEWS FACT BOOK

Gasoline octanes, U.S. service stations, for regular, premium, and unleaded gasoline, annual. NATIONAL PETROLEUM NEWS FACT BOOK

Gasoline prices. See also Motor gasoline, average prices

Gasoline prices, average historical data. PLATT'S OIL PRICE HANDBOOK AND OILMANAC

Gasoline prices, domestic averages. ANNUAL STATISTICAL REVIEW

Gasoline prices, domestic, retail, and state and federal taxes, annual. STANDARD AND POOR'S INDUSTRY SURVEYS

Gasoline prices, retail, at service stations, annual. STANDARD AND POOR'S TRADE AND SECURITIES STATISTICS

Gasoline prices, retail, duty and taxes, by country. INTERNATIONAL PETRO-
LEUM

Gasoline prices, retail, in the United States and other OECD countries in U.S.
cents per gallon. INTERNATIONAL OIL DEVELOPMENTS

Gasoline prices, retail, monthly and annual averages of fifty-five U.S. cities.
STANDARD AND POOR'S TRADE AND SECURITIES STATISTICS

Gasoline prices, retail, regular grade, and taxes, annual. BASIC PETROLEUM
DATA BOOK

Gasoline prices, tank wagon prices, average price to dealers, in selected
cities. PLATT'S OIL PRICE HANDBOOK AND OILMANAC

Gasoline prices, wholesale, tank wagon, annual. STANDARD AND POOR'S
INDUSTRY SURVEYS

Gasoline prices, wholesale price index, annual and monthly. PLATT'S OIL
PRICE HANDBOOK AND OILMANAC

Gasoline prices, world prices (AFM) barge quotations. PLATT'S OIL PRICE
HANDBOOK AND OILMANAC

Gasoline prices at service stations in selected cities. PLATT'S OIL PRICE
HANDBOOK AND OILMANAC

Gasoline prices in selected cities and states. OIL DAILY

Gasoline price trends of regular grade gasoline, annual. ENERGY STATISTICS

Gasoline production. See also Gasoline, domestic production

Gasoline production, imports, exports, consumption, and supply in OECD member countries, annual and quarterly. QUARTERLY OIL STATISTICS

Gasoline sales by grade, by metropolitan area. NATIONAL PETROLEUM
NEWS FACT BOOK

Gasoline service stations. See also Service stations

Gasoline service stations, analysis of customers. NATIONAL PETROLEUM
NEWS FACT BOOK

Gasoline service stations, total number, sales, average sales per station, and
product sales by type, annual. NATIONAL PETROLEUM NEWS FACT BOOK

Gasoline tax, retail, state and federal, annual. STANDARD AND POOR'S
TRADE AND SECURITIES STATISTICS

Gasoline tax rates and collections by state and federal governments. NATIONAL PETROLEUM NEWS FACT BOOK

Gasoline transported through interstate pipelines, annual. STANDARD AND POOR'S TRADE AND SECURITIES STATISTICS

Gas prices, average retail residential prices. GAS FACTS

Gas production in the United States, area proved productive in acres, by state. OIL PRODUCING INDUSTRY IN YOUR STATE

Gas ranges, manufacturers' shipments, annual. GAS FACTS; MOODY'S PUBLIC UTILITY MANUAL

Gas reservoirs, ultimate capacity by state. GAS FACTS

Gas stored and native in storage reservoirs, total volume by state. GAS FACTS

Gas stored in storage reservoirs, total volume by state. GAS FACTS

Gas turbine generators, domestic sales. STANDARD AND POOR'S INDUSTRY SURVEYS

Gas turbine plants, planned or under construction, description, location, scheduled year of completion, and kilowatt capacity. STEAM ELECTRIC PLANT FACTORS

Gas utilities. See also Gas utility industry

Gas utilities, composite financial and related statistics, annual. STANDARD AND POOR'S INDUSTRY SURVEYS

Gas utilities, directory showing name of the company and the city served. MOODY'S PUBLIC UTILITY MANUAL

Gas utilities, Federal Reserve Board indexes of industrial production, monthly and annual averages. STANDARD AND POOR'S TRADE AND SECURITIES STATISTICS

Gas utilities, forecast of selected statistical data and annual growth rates for the United States. PREDICASTS

Gas utilities, forecast of selected statistical data and annual growth rates for the world, by region and country. WORLD CASTS

Gas utilities, hourly wages, weekly earnings, and hours worked, monthly and annual averages. STANDARD AND POOR'S TRADE AND SECURITIES STATISTICS

Gas utilities, income account, balance sheet, financial and operating ratios, capitalization, area served, customers, and kilowatts sold. FINANCIAL STATISTICS OF PUBLIC UTILITIES; MOODY'S PUBLIC UTILITY MANUAL

Gas utilities, new construction expenditures, annual. MOODY'S PUBLIC UTILITY MANUAL; STANDARD AND POOR'S INDUSTRY SURVEYS

Gas utilities, new construction expenditures by type of facility, annual. GAS FACTS

Gas utilities, sales by class of service, annual. MOODY'S PUBLIC UTILITY MANUAL; STANDARD AND POOR'S INDUSTRY SURVEYS

Gas utility industry. See also Gas utilities

Gas utility industry, composite income accounts and balance sheet data, by type, selected years. MOODY'S PUBLIC UTILITY MANUAL; STANDARD AND POOR'S INDUSTRY SURVEYS

Gas utility industry, construction expenditures by type of facility, annual. MOODY'S PUBLIC UTILITY MANUAL; STANDARD AND POOR'S INDUSTRY SURVEYS

Gas utility industry, consumption per customer by state and by residential, commercial, industrial, and other class of service. GAS FACTS

Gas utility industry, customers and sales by class of service, annual. MOODY'S PUBLIC UTILITY MANUAL

Gas utility industry, domestic sales by class of service, annual. MOODY'S PUBLIC UTILITY MANUAL; OIL AND GAS CHEMICAL SERVICE

Gas utility industry, employees and payroll, annual. MOODY'S PUBLIC UTILITY MANUAL

Gas utility industry, employees and payroll data, by type of company, annual. GAS FACTS

Gas utility industry, firm and interruptible gas revenues by state. GAS FACTS

Gas utility industry, firm and interruptible gas sales by state. GAS FACTS

Gas utility industry, gas supply and disposition. GAS FACTS

Gas utility industry, heating saturations and house-heating customers by census division. MOODY'S PUBLIC UTILITY MANUAL

Gas utility industry, house-heating customers and heating saturations by census division and by state. GAS FACTS

Gas utility industry, indexes of common stock prices, dividends, and yields, by type of company. GAS FACTS

Gas utility industry, installed compressor horsepower. GAS FACTS

Gas utility industry, investor-owned. See also Gas utilities

Gas utility industry, investor-owned, composite income accounts and balance sheet, by type of company. GAS FACTS

Gas utility industry, investor-owned, selected composite analytical ratios. GAS FACTS

Gas utility industry, large-volume sales. See also Gas utility industry, sales

Gas utility industry, large-volume sales by type of industry. STANDARD AND POOR'S INDUSTRY SURVEYS

Gas utility industry, large-volume sales by type of industry and by region. GAS FACTS

Gas utility industry, maximum and minimum day send-out by source of supply. GAS FACTS

Gas utility industry, miles of main by type of main. MOODY'S PUBLIC UTILITY MANUAL

Gas utility industry, miles of pipeline and main by type, annual. BASIC PETROLEUM DATA BOOK; ENERGY STATISTICS

Gas utility industry, miles of pipeline and main by type, by state. GAS FACTS

Gas utility industry, miles of pipeline and main installed by type, by state, and by type of pipe. GAS FACTS

Gas utility industry, prices by state and by class of service. GAS FACTS

Gas utility industry, requirements of steel pipe for new construction, maintenance and repair by size of pipe. GAS FACTS

Gas utility industry, revenue by residential, commercial, industrial, and other class of service and by state. GAS FACTS

Gas utility industry, revenues from gas for resale, by state. GAS FACTS

Gas utility industry, revenues of large-volume sales by type of industry and by region. GAS FACTS

Gas utility industry, sales. See also Gas utility industry, large-volume sales

Gas utility industry, sales by month and year. MOODY'S PUBLIC UTILITY MANUAL

Gas utility industry, sales by residential, commercial, industrial, and other class service, by state. GAS FACTS

Gas utility industry, sales, monthly totals. GAS FACTS

Gas utility industry, sales of gas for electric generation, by state. GAS FACTS

Gas utility industry, sales of gas for resale, by state. GAS FACTS

Gas utility industry, sales, revenues, and customers by class of service, annual. STANDARD AND POOR'S TRADE AND SECURITIES STATISTICS

Gas utility industry, security issues, cost and yield by type of issue. GAS FACTS

Gas utility industry, selected analytical financial ratios by type, selected years. MOODY'S PUBLIC UTILITY MANUAL

Gas utility industry plant by type of company and by plant function. GAS FACTS

Gas utility prices by residential, commercial, industrial, and other class of service. GAS FACTS

Gas utility revenues from sales of gas for electric generation, by state. GAS FACTS

Gas water heaters, manufacturers' shipments, by state. GAS FACTS

Gas wells, average cost per foot, by depth classes. GAS FACTS

Gas wells, new completions, abandonments, and total producing wells. OIL AND GAS CHEMICAL SERVICE

Gas wells, drilling and equipping, by depth intervals, estimated costs in the United States. ANNUAL STATISTICAL REVIEW

Gas wells, footage drilled and estimated drilling costs, by state. ANNUAL STATISTICAL REVIEW

Gas wells, new completions by state. GAS FACTS

Gas wells, new field wildcat wells drilled and proportion successful. GAS FACTS

Gas wells, number drilled, annual. STANDARD AND POOR'S TRADE AND SECURITIES STATISTICS

Gas wells, number of producing wells, by state. GAS FACTS

Gas wells, offshore drilling operations and expenditures. GAS FACTS

Gas wells, onshore, estimated drilling costs in the United States, annual. BASIC PETROLEUM DATA BOOK

Gas wells, producing and drilled. TWENTIETH CENTURY PETROLEUM STATISTICS

Gas wells, productive, average annual drilling costs in the United States. ANNUAL STATISTICAL REVIEW

Gas wells, productive, number of wells, footage drilled, and cost of drilling and equipping wells and dry holes, annual. ANNUAL STATISTICAL REVIEW

Gas wells and dry holes drilled as exploratory tests, by state. GAS FACTS

Gas wells drilled, number of producing wells, by state. GAS FACTS

Geographical area served by gas and electric utilities in the United States. FINANCIAL STATISTICS OF PUBLIC UTILITIES; MOODY'S PUBLIC UTILITY MANUAL

Geological and geophysical expenses by world petroleum industry, by select countries. CAPITAL INVESTMENTS IN THE PETROLEUM INDUSTRY

Geophysical activity in the United States (crew months worked), by PAD district. OIL PRODUCING INDUSTRY IN YOUR STATE

Geophysical exploration in the free world by selected countries, annual. BASIC PETROLEUM DATA BOOK

Georgia, footage drilled of new wells. TWENTIETH CENTURY PETROLEUM STATISTICS

Georgia, installed generating capacity, electricity generation by type of prime mover, and electric energy sold by type of customer. EDISON ELECTRIC INSTITUTE STATISTICAL YEARBOOK

Georgia, oil companies' market shares, by company. NATIONAL PETROLEUM NEWS FACT BOOK

Georgia, total oil and gas wells, dry holes, development and exploratory wells drilled, by quarter. QUARTERLY REVIEW OF DRILLING STATISTICS

Geothermal power, consumption in the United States through the year 2000 by consuming sector and as a percent of total energy input. BASIC PETROLEUM DATA BOOK

Geothermal power, world consumption through 1990 and as a percent of total energy consumption. BASIC PETROLEUM DATA BOOK

Germany (East), crude production, percent of proved world reserves of crude and natural gas, tank ship fleet, deadweight tonnage, and tank ships under construction. TWENTIETH CENTURY PETROLEUM STATISTICS

Germany (West), crude production, percent of proved world reserves of crude and natural gas, refining capacity, demand for refined products, tank ship fleet, deadweight tonnage, and tank ships under construction. TWENTIETH CENTURY PETROLEUM STATISTICS

Germany (West), estimated imports of crude oil and refined products from Arab and non-Arab countries, by original crude source. INTERNATIONAL OIL DEVELOPMENTS

Germany (West), imports and exports of crude and products. INTERNATIONAL OIL DEVELOPMENTS

Germany (West), primary energy production, annual. INTERNATIONAL ECONOMIC REPORT OF THE PRESIDENT

Germany (West), production, consumption, and exports and imports by direction, of crude oil and crude oil products by type, and of natural gas and liquids. QUARTERLY OIL STATISTICS

Germany (West), production, consumption, exports, and imports, by function and user, of energy sources by type. STATISTICS OF ENERGY

Getty Oil Company, crude oil prices (posted). PLATT'S OIL PRICE HANDBOOK AND OILMANAC

Ghana, refining capacity and demand for refined products. TWENTIETH CENTURY PETROLEUM STATISTICS

Government-owned electric utilities. See also Electric utilities, publicly owned

Government-owned electric utilities, installed generating capacity, annual. EDISON ELECTRIC INSTITUTE STATISTICAL YEARBOOK

Greece, production, consumption and imports and exports by direction of crude oil and crude oil products by type, and of natural gas and liquids. QUARTERLY OIL STATISTICS

Greece, production, imports, exports, and consumption, by function and user, of energy sources by type of energy. STATISTICS OF ENERGY

Greece, refining capacity and demand for refined products, tank ship fleet, deadweight tonnage, and tank ships under construction. TWENTIETH CENTURY PETROLEUM STATISTICS

Greensboro (N.C.), motor gasoline refinery and terminal prices, regular and premium. PLATT'S OIL PRICE HANDBOOK AND OILMANAC

Guatemala, demand for refined products. TWENTIETH CENTURY PETROLEUM STATISTICS

Gulf Coast gasoline cargoes spot sales and fuel oil cargoes. OIL DAILY

Gulf of Mexico, total oil and gas wells, dry holes, development, and exploratory wells drilled, by quarter. QUARTERLY REVIEW OF OIL STATISTICS

Gulf Oil Company (U.S.), crude oil prices (posted). PLATT'S OIL PRICE HANDBOOK AND OILMANAC

Haulage methods and units used at bituminous coal mines. BITUMINOUS COAL DATA

Haul-loaded conveyers at underground bituminous coal mines. BITUMINOUS COAL DATA

Hawaii, installed generating capacity, electricity generation by type of prime mover, and electric energy sold by type of customer. EDISON ELECTRIC INSTITUTE STATISTICAL YEARBOOK

Hawaii, oil companies' market shares, by company. NATIONAL PETROLEUM NEWS FACT BOOK

Heating degree days. See Degree days

Heating equipment, gas, manufacturers' shipments by type, annual. MOODY'S PUBLIC UTILITY MANUAL

Heating equipment, manufacturers' shipments, annual. GAS FACTS

Heating oil. See also Home heating oil

Heating oil sales in the United States, annual. BASIC PETROLEUM DATA BOOK

Heating saturations in the gas utility industry by census division. MOODY'S PUBLIC UTILITY MANUAL; GAS FACTS

Heat rate, average number of BTUs required to produce one kilowatt-hour, by region. STEAM ELECTRIC PLANT FACTORS

Heat rate in the consumption of fossil fuels for electricity generation. EDISON ELECTRIC INSTITUTE STATISTICAL YEARBOOK

Heat units and other fuels expressed in terms of million tons of oil. BP STATISTICAL REVIEW OF THE WORLD OIL INDUSTRY

Heavy fuel, wholesale prices per gallon, annual. NATIONAL PETROLEUM NEWS FACT BOOK

Heavy-water-moderated gas-cooled and boiling-light-water-cooled reactors, description, and net electrical generation by country. POWER REACTORS IN MEMBER STATES

Holes, drilled. See also Wells; Oil wells; Gas wells

Holes drilled, number and proportion of successful holes, by type of exploratory hole. GAS FACTS

Home heating by type of fuel used. NATIONAL PETROLEUM.NEWS FACT BOOK

Home heating oil. See also Heating oil

Home heating oil, annual average price in current and constant cents. BASIC PETROLEUM DATA BOOK

Hopper cars. See Railroads

Hours and earnings, average weekly, in electric, gas, and sanitary services, annual. MOODY'S PUBLIC UTILITY MANUAL

House-heating customers of the gas utility industry by census division and by state. GAS FACTS

Household energy consumption. See also under specific fuels

Household energy consumption in the United States, annual. BASIC PETROLEUM DATA BOOK

Houston (Tex.), distillates and residual fuel oils, refinery and terminal prices, high, low, and average. PLATT'S OIL PRICE HANDBOOK AND OILMANAC

Hungary, crude production, percent of world proved reserves of crude and natural gas, and refining capacity. TWENTIETH CENTURY PETROLEUM STATISTICS

Hungary, oil and natural gas production, consumption, and trade. INTERNATIONAL OIL DEVELOPMENTS

Hydroelectric plants, planned or under construction, description, location, scheduled year of completion, and kilowatt capacity. STEAM ELECTRIC PLANT FACTORS

Hydroelectric power, domestic consumption. MONTHLY ENERGY REVIEW

Hydroelectric power, domestic consumption by household, commercial, industrial, transportation, and electricity generation users, annual. GAS FACTS

Hydroelectric power, domestic consumption through the year 2000 by consuming sector and as a percent of gross energy input. BASIC PETROLEUM DATA BOOK

Hydroelectric power, free-world demand, annual. STANDARD AND POOR'S INDUSTRY SURVEYS

Hydroelectric power, production in the United States and as a percent of total energy production, annual. BASIC PETROLEUM DATA BOOK

Hydroelectric power, world consumption through 1990 and as a percent of total energy consumption. BASIC PETROLEUM DATA BOOK

Hydroelectric power-generating plants, kilowatts generated and percent of total electricity generated. STANDARD AND POOR'S INDUSTRY SURVEYS

Hydroelectric utilities, electricity generation by type of ownership, annual. EDISON ELECTRIC INSTITUTE STATISTICAL YEARBOOK

Hydroelectric utilities, reported new capability addition, by type of ownership. MOODY'S PUBLIC UTILITY MANUAL

Iceland, production, consumption, exports, and imports, by function and user, of energy sources by type. STATISTICS OF ENERGY

Iceland, production, consumption, exports and imports by direction, of crude oil and crude oil products, and of natural gas and liquids. QUARTERLY OIL STATISTICS

Idaho, installed generating capacity, electricity generation by type of prime mover, and electric energy sold by type of customer. EDISON ELECTRIC INSTITUTE STATISTICAL YEARBOOK

Idaho, oil companies' market shares, by company. NATIONAL PETROLEUM NEWS FACT BOOK

Illinois, crude production, average daily production per well, footage drilled, marketed production of natural gas, reserves of natural gas and crude, well completions, and producing oil wells. TWENTIETH CENTURY PETROLEUM STATISTICS

Illinois, installed generating capacity, electricity generation by type of prime mover, and electric energy sold by type of customer. EDISON ELECTRIC INSTITUTE STATISTICAL YEARBOOK

Illinois, oil and natural gas production, reserves, exploration, development, prices, and related data, annual. OIL PRODUCING INDUSTRY IN YOUR STATE

Illinois, oil companies' market shares, by company. NATIONAL PETROLEUM NEWS FACT BOOK

Illinois, total oil and gas wells, dry holes, development and exploratory wells drilled, by quarter. QUARTERLY REVIEW OF DRILLING STATISTICS

Imperial Oil Limited, crude oil prices (posted). PLATT'S OIL PRICE HANDBOOK AND OILMANAC

Imports (domestic) as percent of production and demand for petroleum, annual. BASIC PETROLEUM DATA BOOK

Imports from OPEC countries by the developed countries, by country, total value. INTERNATIONAL OIL DEVELOPMENTS

Imports of Arab and non-Arab oil by United States, Japan, European Community, United Kingdom, France, West Germany, and Italy. INTERNATIONAL ECONOMIC REPORT OF THE PRESIDENT

Imports of crude by source, total and daily averages, fuel oil, kerosene, motor fuel, and refined products. TWENTIETH CENTURY PETROLEUM STATISTICS

Imports of electricity from Canada and Mexico, annual. EDISON ELECTRIC INSTITUTE STATISTICAL YEARBOOK

Imports of refined products by selected countries, from Arab and non-Arab sources. INTERNATIONAL ECONOMIC REPORT OF THE PRESIDENT

Income account of gas and electric companies in the United States. FINANCIAL STATISTICS OF PUBLIC UTILITIES; PUBLIC UTILITY MANUAL; OIL AND GAS CHEMICAL SERVICE

Income (gross and net), of gas and electric utilities in the United States. FINANCIAL STATISTICS OF PUBLIC UTILITIES; MOODY'S PUBLIC UTILITY MANUAL; OIL AND GAS CHEMICAL SERVICE

Income statement of the petroleum industry. FINANCIAL ANAYSIS OF A GROUP OF PETROLEUM COMPANIES

Index of domestic retail gasoline prices compared with the consumer price index, historical data, annual. ANNUAL STATISTICAL REVIEW

Index of proved fuel reserves, domestic. TWENTIETH CENTURY PETROLEUM STATISTICS

Index of weekly electric output, all electric utility industry, annual. EDISON ELECTRIC INSTITUTE STATISTICAL YEARBOOK

India, crude production, total and as a percent of world production, proved reserves of crude and natural gas, demand for refined products, refining capacity, tank ship fleet, and deadweight tonnage. TWENTIETH CENTURY PETROLEUM STATISTICS

Indiana, crude production, daily average production per well, marketed production of natural gas, reserves of natural gas and crude, well completions, and producing oil wells. TWENTIETH CENTURY PETROLEUM STATISTICS

Indiana, installed generating capacity, electricity generation by type of prime mover, and electric energy sold by type of customer. EDISON ELECTRIC INSTITUTE STATISTICAL YEARBOOK

Indiana, oil and natural gas production, reserves, exploration, developments, price, related data, annual. OIL PRODUCING INDUSTRY IN YOUR STATE

Indiana oil companies' market shares, by company. NATIONAL PETROLEUM NEWS FACT BOOK

Indiana, total oil and gas wells, dry holes, development, exploratory wells drilled, by quarter. QUARTERLY REVIEW OF DRILLING STATISTICS

Indianapolis (Ind.), gasoline and fuel oil prices. OIL DAILY

Indonesia, crude oil production in barrels per day. INTERNATIONAL OIL DEVELOPMENTS

Indonesia, crude oil production in barrels per day and as a percent of total production. INTERNATIONAL ECONOMIC REPORT OF THE PRESIDENT

Indonesia, crude production, total and as a percent of world production, exports to the United States, reserves of crude and natural gas, refining capacity, refined products demand, and tank ship fleet. TWENTIETH CENTURY PETROLEUM STATISTICS

Indonesia, imports from and exports to the United States and other developed countries, total value. INTERNATIONAL OIL DEVELOPMENTS

Indonesia, natural gas liquid production in barrels per day. INTERNATIONAL OIL DEVELOPMENTS

Indonesia, oil reserves and production, exports of crude oil and petroleum to the United States, annual. BASIC PETROLEUM DATA BOOK

Industrial customers, sales of electric power, annual. STANDARD AND POOR'S INDUSTRY SURVEYS

Industrial electricity sales revenue, annual. EDISON ELECTRIC INSTITUTE STATISTICAL YEARBOOK

Industrial energy, domestic consumption. MONTHLY ENERGY REVIEW

Industrial energy, domestic consumption, annual. BASIC PETROLEUM DATA BOOK

Industrial sales of electricity, annual. EDISON ELECTRIC INSTITUTE STATISTI-
CAL YEARBOOK

Injuries. See also Accidents

Injuries, fatal and nonfatal at bituminous coal mines, annual. BITUMINOUS
COAL DATA

Injuries, rates in the bituminous coal industry. COAL FACTS

Injury (disability), of employees, frequency and severity rates, by type of gas.
GAS FACTS

Installed generating capacity. See Electric utilities, installed capacity of
generating plants; Steam-electric utilities, installed generating capacity

Interest on long-term debt of gas and electric utilities in the United States.
FINANCIAL STATISTICS OF PUBLIC UTILITIES; MOODY'S PUBLIC UTILITY
MANUAL

Internal combustion electric utilities, electricity generation by type of ownership, annual. EDISON ELECTRIC INSTITUTE STATISTICAL YEARBOOK

Internal combustion electric utilities, installed generating capacity by ownership, annual. EDISON ELECTRIC INSTITUTE STATISTICAL YEARBOOK

Internal combustion plants, kilowatts generated and as a percent of total
production, annual. STANDARD AND POOR'S INDUSTRY SURVEYS

Internal combustion plants, planned or under construction, description, location,
scheduled year of completion and kilowatt capacity. STEAM ELECTRIC PLANT
FACTORS

Internal combustion plants, reported new capability addition, by type of ownership. MOODY'S PUBLIC UTILITY MANUAL

Interstate pipeline transportation of crude and refined oil and gasoline, annual.
STANDARD AND POOR'S TRADE AND SECURITIES STATISTICS

Investments (gross and net), in fixed assets, of the world petroleum industry,
by select countries. CAPITAL INVESTMENTS IN THE PETROLEUM INDUSTRY

Investor-owned electric utilities. See also Electric utilities, investor-owned;
Electric utilities, privately owned

Investor-owned electric utilities, composite balance sheets and income statements, annual. EDISON ELECTRIC INSTITUTE STATISTICAL YEARBOOK

Investor-owned electric utilities, installed generating capacity, annual. EDI-
SON ELECTRIC INSTITUTE STATISTICAL YEARBOOK

Investor-owned gas utilities. See also Gas utility industry, investor-owned

Iowa, gasoline and fuel oil prices. OIL DAILY

Iowa, installed generating capacity, electricity generation by type of prime mover, and electric energy sold by type of customer. EDISON ELECTRIC INSTITUTE STATISTICAL YEARBOOK

Iowa, new wells, footage drilled. TWENTIETH CENTURY PETROLEUM STATISTICS

Iowa, oil companies' market shares, by company. NATIONAL PETROLEUM NEWS FACT BOOK

Iran, crude oil prices by oil field. PLATT'S OIL PRICE HANDBOOK AND OIL-MANAC

Iran, crude oil production barrels per day. INTERNATIONAL OIL DEVELOPMENTS

Iran, crude oil production in barrels per day and as a percent of total production. INTERNATIONAL ECONOMIC REPORT OF THE PRESIDENT

Iran, crude production, total and as a percent of world production, reserves of crude and natural gas, posted price of crude per barrel, refining capacity, refined products demand, export to the United States, tank ship fleet, deadweight tonnage, and construction. TWENTIETH CENTURY PETROLEUM STATISTICS

Iran, imports from and exports to the United States and other developed countries, total value. INTERNATIONAL OIL DEVELOPMENTS

Iran, natural gas liquids production in barrels per day. INTERNATIONAL OIL DEVELOPMENTS

Iran, oil reserves and production, exports of crude oil and petroleum to the United States, annual. BASIC PETROLEUM DATA BOOK

Iraq, crude oil prices by oil field. PLATT'S OIL PRICE HANDBOOK AND OILMANAC

Iraq, crude oil production in barrels per day. INTERNATIONAL OIL DEVELOPMENTS

Iraq, crude oil production in barrels per day and as a percent of total production. INTERNATIONAL ECONOMIC REPORT OF THE PRESIDENT

Iraq, crude production, total and as a percent of world production, exports to the United States, reserves of crude and natural gas, posted price of crude per barrel, refined products demand, refining capacity, and tank ships. TWENTIETH CENTURY PETROLEUM STATISTICS

Iraq, imports from and exports to the United States and other developed countries, total value. INTERNATIONAL OIL DEVELOPMENTS

Iraq, oil reserves and production, exports of crude oil and petroleum to the United States, annual. BASIC PETROLEUM DATA BOOK

Ireland, production, consumption, exports, and imports by function and user, of energy sources by type. STATISTICS OF ENERGY

Ireland, production, consumption, exports and imports by direction, of crude oil and crude oil products by type, and of natural gas and liquids. QUARTERLY OIL STATISTICS

Ireland, refining capacity and refined products demand. TWENTIETH CENTURY PETROLEUM STATISTICS

Israel, crude production, reserves of crude and natural gas, refining capacity, and tank ship fleet. TWENTIETH CENTURY PETROLEUM STATISTICS

Italy, crude production, total and as a percent of world production, reserves of crude and natural gas, refining capacity and refined products demand. TWENTIETH CENTURY PETROLEUM STATISTICS

Italy, estimated imports of crude oil and refined products from Arab and non-Arab countries by original crude source. INTERNATIONAL OIL DEVELOPMENTS

Italy, imports and exports of crude and products. INTERNATIONAL OIL DEVELOPMENTS

Italy, primary energy production, annual. INTERNATIONAL ECONOMIC REPORT OF THE PRESIDENT

Italy, production, consumption, exports and imports by direction, of crude oil and crude oil products by type, and of natural gas and liquids. QUARTERLY OIL STATISTICS

Italy, production, imports, exports, and consumption, by function and user, of energy sources by type. STATISTICS OF ENERGY

Jacksonville (Fla.), distillates and residual fuel oils refinery and terminal prices, high, low, and average. PLATT'S OIL PRICE HANDBOOK AND OILMANAC

Jamaica, refining capacity, refined products demand, and tank ship fleet. TWENTIETH CENTURY PETROLEUM STATISTICS

Japan, crude production, reserves of crude and natural gas, refining capacity,

refined products demand, and tank ship fleet. TWENTIETH CENTURY PETRO-
LEUM STATISTICS

Japan, estimated imports of crude oil and refined products from Arab and non-
Arab countries, by original crude source. INTERNATIONAL OIL DEVELOP-
MENTS

Japan, imports and exports of crude and products. INTERNATIONAL OIL
DEVELOPMENTS

Japan, primary energy production, annual. INTERNATIONAL ECONOMIC
REPORT OF THE PRESIDENT

Japan, production, consumption, exports, and imports, by function and user,
of energy sources by type. STATISTICS OF ENERGY

Japan, production, consumption, exports and imports by direction, of crude
oil and crude oil products, and of natural gas and liquids. QUARTERLY OIL
STATISTICS

Japanese consumption of Middle East oil, annual. BASIC PETROLEUM DATA
BOOK

Jet fuel, domestic consumption. MONTHLY ENERGY REVIEW

Jet fuel, domestic demand, annual. TWENTIETH CENTURY PETROLEUM
STATISTICS

Jet fuel, domestic demand, production, and refinery yield. TWENTIETH
CENTURY PETROLEUM STATISTICS

Jet fuel, domestic production, annual. STANDARD AND POOR'S TRADE
AND SECURITIES STATISTICS

Jet fuel, domestic production and stocks, annual. STANDARD AND POOR'S
TRADE AND SECURITIES STATISTICS

Jet fuel, domestic shipments to PAD districts. ENERGY STATISTICS

Jet fuel, domestic shipments to PAD districts, by use, daily averages. AN-
NUAL STATISTICAL REVIEW

Jet fuel, forecast of selected statistical data and annual growth rates for the
United States. PREDICASTS

Jet fuel, forecast of selected statistical data and annual growth rates for the
world, by region and country. WORLD CASTS

Jet fuel, kerosene type, domestic supply and demand, daily averages. AN-
NUAL STATISTICAL REVIEW

Jet fuel, naphtha and kerosene type, domestic supply and demand, annual. BASIC PETROLEUM DATA BOOK

Jet fuel, naphtha type, domestic supply and demand, daily averages. ANNUAL STATISTICAL REVIEW

Jet fuel, production, imports, exports, and total internal consumption in OECD member countries, annual. STATISTICS OF ENERGY

Jet fuel, production, stocks, imports, exports, and domestic demand in the United States, annual. ENERGY STATISTICS

Jet fuel, production, trade, and total and per capita apparent consumption, by country, annual. WORLD ENERGY SUPPLIES

Jet fuel, production in the free world, percentages. INTERNATIONAL PETROLEUM

Jet operating expenses including fuel and oil costs by airline and type. ENERGY STATISTICS

Jobbers and commission agents, for gasoline distribution, total number, by company. NATIONAL PETROLEUM NEWS FACT BOOK

Jordan, refining capacity. TWENTIETH CENTURY PETROLEUM STATISTICS

Kansas, crude production; new wells drilled; average production per well; marketed production of natural gas; reserves of crude, natural gas, and natural gas liquids. TWENTIETH CENTURY PETROLEUM STATISTICS

Kansas, gasoline and fuel oil prices. OIL DAILY

Kansas, installed generating capacity, electricity generation by type of prime mover, and electric energy sold by type of customer. EDISON ELECTRIC INSTITUTE STATISTICAL YEARBOOK

Kansas, oil and natural gas production, reserves, exploration, development, prices, and related data, annual. OIL PRODUCING INDUSTRY IN YOUR STATE

Kansas, oil companies' market shares, by company. NATIONAL PETROLEUM NEWS FACT BOOK

Kansas, total oil and gas wells, dry holes, development, and exploratory wells drilled, by quarter. QUARTERLY REVIEW OF DRILLING STATISTICS

Kentucky, crude production; new wells drilled; daily average production per well; marketed production of natural gas; reserves of natural gas, crude and natural gas liquids. TWENTIETH CENTURY PETROLEUM STATISTICS

Kentucky, installed generating capacity, electricity generation by type of

prime mover, and electric energy sold by type of prime mover. EDISON ELEC-
TRIC INSTITUTE STATISTICAL YEARBOOK

Kentucky, oil and natural gas production, reserves, exploration, development,
prices, and related data, annual. OIL PRODUCING INDUSTRY IN YOUR
STATE

Kentucky, oil companies' market shares, by company. NATIONAL PETROLEUM
NEWS FACT BOOK

Kentucky, total oil and gas wells, dry holes, development, and exploratory
wells drilled, by quarter. QUARTERLY REVIEW OF DRILLING STATISTICS

Kenya, refining capacity and refined products demand. TWENTIETH CENTURY
PETROLEUM STATISTICS

Kerosene, average wholesale prices, historical data. PLATT'S OIL PRICE HAND-
BOOK AND OILMANAC

Kerosene, contribution to freight revenue of U.S. railroads, annual. ENERGY
STATISTICS

Kerosene, conversion factors. BP STATISTICAL REVIEW OF THE WORLD OIL
INDUSTRY

Kerosene, domestic demand, annual. NATIONAL PETROLEUM NEWS FACT
BOOK; STANDARD AND POOR'S INDUSTRY SURVEYS; TWENTIETH CENTURY
PETROLEUM STATISTICS

Kerosene, domestic demand and supply, annual. BASIC PETROLEUM DATA
BOOK

Kerosene, domestic demand by use, and supply, daily averages. ANNUAL
STATISTICAL REVIEW

Kerosene, domestic production, annual. STANDARD AND POOR'S TRADE AND
SECURITIES STATISTICS

Kerosene, domestic production, consumption, imports, exports, and stocks,
annual. STANDARD AND POOR'S TRADE AND SECURITIES STATISTICS

Kerosene, free-world production, percentages. INTERNATIONAL PETROLEUM

Kerosene, jet grade, European bulk prices. PLATT'S OIL PRICE HANDBOOK
AND OILMANAC

Kerosene, jet grade and water white, prices for Gulf Coast cargoes, high, low,
and average. PLATT'S OIL PRICE HANDBOOK AND OILMANAC

Kerosene, production, exports, imports, consumption, and supply in OECD countries, annual and quarterly. QUARTERLY OIL STATISTICS

Kerosene, production, imports, exports, and total internal consumption in OECD member countries, annual. STATISTICS OF ENERGY

Kerosene, production, imports, exports, demand, stocks, and refinery yield. TWENTIETH CENTURY PETROLEUM STATISTICS

Kerosene, production, trade, and total per capita apparent consumption, by country, annual. WORLD ENERGY SUPPLIES

Kerosene, retail prices, including duty and taxes, by country. INTERNATIONAL PETROLEUM

Kerosene, wholesale prices per gallon, annual. NATIONAL PETROLEUM NEWS FACT BOOK

Korea (South), demand for refined products and tank ship fleet. TWENTIETH CENTURY PETROLEUM STATISTICS

Kuwait, crude oil prices by oil field. PLATT'S OIL PRICE HANDBOOK AND OILMANAC

Kuwait, crude oil production in barrels per day. INTERNATIONAL OIL DEVELOPMENTS

Kuwait, crude oil production in barrels per day and as a percent of total production. INTERNATIONAL ECONOMIC REPORT OF THE PRESIDENT

Kuwait, crude production, total and as a percent of world production, exports to the United States, reserves of crude and natural gas, posted price of crude per barrel, refining capacity, refined products demand, and tank ship fleet. TWENTIETH CENTURY PETROLEUM STATISTICS

Kuwait, imports from and exports to the United States and other developed countries, total value. INTERNATIONAL OIL DEVELOPMENTS

Kuwait, natural gas liquids production in barrels per day. INTERNATIONAL OIL DEVELOPMENTS

Kuwait, oil reserves and production, and exports to the United States of crude oil and petroleum, annual. BASIC PETROLEUM DATA BOOK

Land area productive of oil or gas, leased for production, by state. OIL PRODUCING INDUSTRY IN YOUR STATE

Lease acquisition, expenditures, on- and offshore, by domestic petroleum industry. CAPITAL INVESTMENT IN THE PETROLEUM INDUSTRY

Lease condensate, total U.S. supply and demand, annual. BASIC PETROLEUM DATA BOOK

Leased area under oil or gas production, by state. OIL PRODUCING INDUS-TRY IN YOUR STATE

Lebanon, refining capacity and refined products demand. TWENTIETH CEN-TURY PETROLEUM STATISTICS

Leonard Crude Oil Company, crude oil prices (posted). PLATT'S OIL PRICE HANDBOOK AND OILMANAC

Liabilities of gas and electric utilities in the United States. FINANCIAL STATISTICS OF PUBLIC UTILITIES; MOODY'S PUBLIC UTILITY MANUAL; OIL AND GAS CHEMICAL SERVICE

Liberia, refined products demand and tank ship fleet. TWENTIETH CENTURY PETROLEUM STATISTICS

Libya, crude oil prices by oil field. PLATT'S OIL PRICE HANDBOOK AND OILMANAC

Libya, crude oil production in barrels per day. INTERNATIONAL OIL DE-VELOPMENTS

Libya, crude oil production in barrels per day and as a percent of total production. INTERNATIONAL ECONOMIC REPORT OF THE PRESIDENT

Libya, imports from and exports to the United States and other developed countries, total value. INTERNATIONAL ECONOMIC DEVELOPMENTS

Libya, natural gas liquids production in barrels per day. INTERNATIONAL OIL DEVELOPMENTS

Libya, oil reserves and production, and exports of crude oil and petroleum to the United States, annual. BASIC PETROLEUM DATA BOOK

Light, power, and gas stocks, new issues, Moody's weighted averages of yields, annual. MOODY'S PUBLIC UTILITY MANUAL

Light fuel, wholesale prices per gallon, annual. NATIONAL PETROLEUM NEWS FACT BOOK

Light-water-cooled and graphite-moderated reactors, description and net electricity generation by country. POWER REACTORS IN MEMBER STATES

Lignite. See also Bituminous coal and lignite; Coal

Lignite, domestic production and consumption. TWENTIETH CENTURY PETRO-LEUM STATISTICS

Lignite, domestic production and consumption in BTUs. TWENTIETH CENTURY PETROLEUM STATISTICS

Lignite, domestic reserves, estimates by state. WORLD COAL TRADE

Lignite and brown coal reserves of the world, estimates by country and region. WORLD COAL TRADE

Lignite coal consumption by electric utilities, coke ovens, steel mills, manufacturing companies, and retail sales, annual. STANDARD AND POOR'S TRADE AND SECURITIES STATISTICS

Lignite mining statistics by state, by type of mine. BITUMINOUS COAL DATA

Line pipe carbon, wholesale price index, annual. BASIC PETROLEUM DATA BOOK

Liquefied gas, production, exports, imports, consumption by function and user, and consumption for transformation in OECD member countries, annual. STATISTICS OF ENERGY

Liquefied gases, total production in the United States, annual. STANDARD AND POOR'S TRADE AND SECURITIES STATISTICS

Liquefied natural gas, forecast of selected statistical data and annual growth rates for the United States. PREDICASTS

Liquefied natural gas, forecast of selected statistical data and annual growth rates for the world by region and country. WORLD CASTS

Liquefied natural gas storage, domestic operations. GAS FACTS

Liquefied petroleum gas, contribution to freight revenue of U.S. railroads, annual. ENERGY STATISTICS

Liquefied petroleum gas, domestic consumption estimates by state. OIL PRODUCING INDUSTRY IN YOUR STATE

Liquefied petroleum gas, domestic demand, annual. NATIONAL PETROLEUM NEWS FACT BOOK

Liquefied petroleum gas, domestic sales, by use, daily averages. ANNUAL STATISTICAL REVIEW

Liquefied petroleum gas, domestic sales by major market. NATIONAL PETROLEUM NEWS FACT BOOK

Liquefied petroleum gas, domestic sales in liquid form to ultimate consumers by gas utility, residential, commercial, industrial, refinery fuel, chemical manufacturing, and synthetic rubber production users. GAS FACTS

Liquefied petroleum gas, prices in selected cities, high, low, and average. PLATT'S OIL PRICE HANDBOOK AND OILMANAC

Liquefied petroleum gas, production, imports, exports, consumption, and stocks in OECD countries, annual and quarterly. QUARTERLY OIL STATISTICS

Liquefied petroleum gas, production, trade, and total and per capital apparent consumption by country, annual. WORLD ENERGY SUPPLIES

Liquefied petroleum gas bulk plants, and terminals, total number, sales, and employees. NATIONAL PETROLEUM NEWS FACT BOOK

Liquefied petroleum gas dealers, total number and sales by state and region. NATIONAL PETROLEUM NEWS FACT BOOK

Liquefied refinery gas, domestic demand, production, and stocks, annual. STANDARD AND POOR'S TRADE AND SECURITIES STATISTICS

Liquefied refinery gas, domestic demand and supply, annual. BASIC PETRO-LEUM DATA BOOK

Liquefied refinery gas, forecast of selected statistical data and annual growth rates for the United States. PREDICASTS

Liquefied refinery gas, forecast of selected statistical data and annual growth rates for the world by region and country. WORLD CASTS

Liquid hydrocarbons, estimated domestic proved recoverable reserves, total and by state. GAS FACTS

Liquid hydrocarbons, proved domestic, reserves, total and annual. ANNUAL STATISTICAL REVIEW

Long-term financing by public utilities (new capital, refunding, and divestment of common and preferred stocks and debt captial). EDISON ELECTRIC INSTITUTE STATISTICAL YEARBOOK

Long wall mining machine production at underground bituminous coal mines. BITUMINOUS COAL DATA

Los Angeles (Calif.), motor gasoline refinery and terminal oil prices, regular and premium. PLATT'S OIL PRICE HANDBOOK AND OILMANAC

Los Angeles (Calif.), propane refinery and terminal prices, high, low, and average. PLATT'S OIL PRICE HANDBOOK AND OILMANAC

Los Angeles (Calif.), refinery and terminal prices of distillates and residual fuel oils. PLATT'S OIL PRICE HANDBOOK AND OILMANAC

Louisiana, crude production, new well completions, daily average production per well, natural gas and liquids production, and reserves of crude and natural gas. TWENTIETH CENTURY PETROLEUM STATISTICS

Louisiana, installed generating capacity, electricity generation by type of prime mover, and electric energy sold by type of prime mover. EDISON ELECTRIC INSTITUTE STATISTICAL YEARBOOK

Louisiana, oil and natural gas production, reserves, exploration, development, prices, and related data, annual. OIL PRODUCING INDUSTRY IN YOUR STATE

Louisiana, oil companies' market shares, by company. NATIONAL PETRO-LEUM NEWS FACT BOOK

Louisiana, total oil and gas wells, dry holes, development, and exploratory wells drilled, by quarter, by district. QUARTERLY REVIEW OF DRILLING STATISTICS

Lubes, domestic demand, annual. NATIONAL PETROLEUM NEWS FACT BOOK

Lubes, refinery and terminal prices, select regions. PLATT'S OIL PRICE HANDBOOK AND OILMANAC

Lubricants, demand, production, and percent refinery yield. TWENTIETH CENTURY PETROLEUM STATISTICS

Lubricants, domestic consumption, production, stocks, exports, and imports, annual. STANDARD AND POOR'S INDUSTRY SURVEYS

Lubricants, domestic demand, annual. TWENTIETH CENTURY PETROLEUM STATISTICS

Lubricants, domestic production, annual. STANDARD AND POOR'S TRADE AND SECURITIES STATISTICS

Lubricants, domestic supply and demand, annual. BASIC PETROLEUM DATA BOOK

Lubricants, domestic supply and demand, by use, daily averages. ANNUAL STATISTICAL REVIEW

Lubricants, free-world production percentages. INTERNATIONAL PETROLEUM

Lubricating oil, world production by country, annual. WORLD ENERGY SUPPLIES

Lubricating oil prices, monthly and annual averages. STANDARD AND POOR'S TRADE AND SECURITIES STATISTICS

Lubricating oils, retail prices, including duty and taxes, by country. INTER-NATIONAL PETROLEUM

Lubricating oils and greases, contribution to freight revenue of U.S. railroads, annual. ENERGY STATISTICS

Lubrication oil materials, wholesale price index, annual and monthly. PLATT'S OIL PRICE HANDBOOK AND OILMANAC

Luxembourg, production, consumption, exports, and imports, by function and user, of energy sources by type. STATISTICS OF ENERGY

Maine, installed generating capacity, electricity generation by type of prime mover, and electric energy sold by type of customer. EDISON ELECTRIC INSTITUTE STATISTICAL YEARBOOK

Maine, oil companies' market shares, by company. NATIONAL PETROLEUM NEWS FACT BOOK

Malagasy, tank ship fleet under construction and in deadweight tonnage. TWENTIETH CENTURY PETROLEUM STATISTICS

Malaysia, crude production, total and as a percent of world production, reserves of crude and natural gas, refining capacity, and tank ship fleet. TWENTIETH CENTURY PETROLEUM STATISTICS

Malaysia/Brunei, crude oil production in barrels per day. INTERNATIONAL OIL DEVELOPMENTS

Malta, tank ship fleet in deadweight tonnage. TWENTIETH CENTURY PETRO-LEUM STATISTICS

Manitoba, crude oil prices by oil field. PLATT'S OIL PRICE HANDBOOK AND OILMANAC

Maps of area served by utility companies. MOODY'S PUBLIC UTILITY MANUAL

Marathon Oil Company, crude oil prices (posted). PLATT'S OIL PRICE HAND-BOOK AND OILMANAC

Marine construction companies, estimated percentage sales to the petroleum industry, and related financial data. STANDARD AND POOR'S INDUSTRY SURVEYS

Marketing, capital and exploration expenditures and gross and net fixed invest-ments, by world petroleum industry, by select countries. CAPITAL INVEST-MENTS IN THE PETROLEUM INDUSTRY

Market shares of oil companies, nationally and by state, by company. NA-
TIONAL PETROLEUM NEWS FACT BOOK

Maryland, footage of new wells drilled. TWENTIETH CENTURY PETROLEUM
STATISTICS

Maryland, installed generating capacity, electricity generation by type of
prime mover, and electric energy sold by type of customer. EDISON ELECTRIC
INSTITUTE STATISTICAL YEARBOOK

Maryland, oil and natural gas production, reserves, exploration, development,
prices, and related data, annual. OIL PRODUCING INDUSTRY IN YOUR
STATE

Maryland, oil companies' market shares, by company. NATIONAL PETROLEUM
NEWS FACT BOOK

Maryland, total oil and gas wells, dry holes, development, and exploratory
wells drilled, by quarter. QUARTERLY REVIEW OF DRILLING STATISTICS

Massachusetts, installed generating capacity, electricity generation by type of
prime mover, and electric energy sold by type of customer. EDISON ELECTRIC
INSTITUTE STATISTICAL YEARBOOK

Massachusetts, oil companies' market shares, by company. NATIONAL PETRO-
LEUM NEWS FACT BOOK

Mechanical loading machines at underground bituminous coal machines. BITU-
MINOUS COAL DATA

Mexico, crude oil production, total and as a percent of world production,
exports to the United States, reserves of crude and natural gas, refined products
demand, refining capacity, and tank ship fleet. TWENTIETH CENTURY PETRO-
LEUM STATISTICS

Mexico, crude oil production in barrels per day. INTERNATIONAL OIL
DEVELOPMENTS

Mexico, crude oil production in barrels per day and as a percent of total
production. INTERNATIONAL ECONOMIC REPORT OF THE PRESIDENT

Mexico, natural gas liquids production in barrels per day. INTERNATIONAL
OIL DEVELOPMENTS

Miami (Fla.), distillates and residue fuel oils refinery and terminal prices,
high, low, and average. PLATT'S OIL PRICE HANDBOOK AND OILMANAC

Michigan, crude production, new wells drilled and average production per
well, natural gas and liquids production, reserves of crude and natural gas.
TWENTIETH CENTURY PETROLEUM STATISTICS

Michigan, gasoline and fuel oil prices. OIL DAILY

Michigan, installed generating capacity, electricity generation by type of prime mover, and electric energy sold by type of customer. EDISON ELECTRIC INSTITUTE STATISTICAL YEARBOOK

Michigan, oil and natural gas production, reserves, exploration, development, prices, and related data, annual. OIL PRODUCING INDUSTRY IN YOUR STATE

Michigan, oil companies' market shares, by company. NATIONAL PETROLEUM NEWS FACT BOOK

Michigan, total oil and gas wells, dry holes, development, and exploratory wells drilled, by quarter. QUARTERLY REVIEW OF DRILLING STATISTICS

Middle distillates. See Distillates (middle)

Middle East, capital and exploration expenditures, and gross and net investments in fixed assets, by domestic petroleum industry. CAPITAL INVESTMENTS IN THE PETROLEUM INDUSTRY

Middle East, crude oil production in barrels per day and as a percent of total production. INTERNATIONAL ECONOMIC REPORT OF THE PRESIDENT

Middle East, crude production by countries, exports to the United States, production demand ratio, reserves of crude and natural gas, refining capacity, refined products demand, and tank ship fleet. TWENTIETH CENTURY PETROLEUM STATISTICS

Middle East, exports of crude production to other countries by country, annual. WORLD ENERGY SUPPLIES

Middle East, exports of oil to selected areas, annual. BASIC PETROLEUM DATA BOOK

Milwaukee (Wis.), gasoline and fuel oil prices. OIL DAILY

Mineral energy fuels, domestic demand, annual. TWENTIETH CENTURY PETROLEUM STATISTICS

Mineral energy fuels, domestic production and consumption in BTUs. TWENTIETH CENTURY PETROLEUM STATISTICS

Mineral energy fuels and electricity, production and consumption from water and nuclear power, annual. BITUMINOUS COAL DATA

Mineral energy resources, domestic consumption, annual. GAS FACTS

Mineral energy resources, domestic gross consumption, annual. BASIC PETROLEUM DATA BOOK

Mineral energy resources, domestic production, annual. BASIC PETROLEUM DATA BOOK; GAS FACTS

Mineral energy resources, domestic production and electricity from hydro and nuclear power, daily averages. ANNUAL STATISTICAL REVIEW

Mineral energy resources and electricity from hydro and nuclear power, domestic, calculated consumption, daily averages. ANNUAL STATISTICAL REVIEW

Mineral fuels, domestic production, nationally and by state. COAL FACTS

Mines, bituminous coal and lignite, number and production; by state, size of output, and type of mining. ENERGY STATISTICS

Mining methods at underground bituminous coal mines. BITUMINOUS COAL DATA

Minneapolis/St. Paul (Minn.), motor gasoline refinery and terminal prices, regular and premium. PLATT'S OIL PRICE HANDBOOK AND OILMANAC

Minneapolis/St. Paul (Minn.), refinery and terminal prices of distillates and residual fuel oils. PLATT'S OIL PRICE HANDBOOK AND OILMANAC

Minnesota, gasoline and fuel oil prices. OIL DAILY

Minnesota, installed generating capacity, electricity generation by type of prime mover, and electric energy sold by type of customer. EDISON ELECTRIC INSTITUTE STATISTICAL YEARBOOK

Minnesota, oil companies' market shares, by company. NATIONAL PETROLEUM NEWS FACT BOOK

Mississippi, crude production; new wells completed and average production per well; natural gas and liquids production; and reserves of crude, natural gas, and liquids. TWENTIETH CENTURY PETROLEUM STATISTICS

Mississippi, installed generating capacity, electricity generation by type of prime mover, and electric energy sold by type of customer. EDISON ELECTRIC INSTITUTE STATISTICAL YEARBOOK

Mississippi, oil and natural gas production, reserves, exploration, developments, prices, and related data, annual. OIL PRODUCING INDUSTRY IN YOUR STATE

Mississippi, oil companies' market shares, by company. NATIONAL PETROLEUM NEWS FACT BOOK

Mississippi, oil wells, gas wells, dry holes, development, and exploratory wells drilled, by quarter. QUARTERLY REVIEW OF DRILLING STATISTICS

Missouri, crude production, new wells drilled, and daily average production per well. TWENTIETH CENTURY PETROLEUM STATISTICS

Missouri, gasoline and fuel oil prices. OIL DAILY

Missouri, installed generating capacity, electricity generation by type of prime mover, and electric energy sold by type of customer. EDISON ELECTRIC INSTITUTE STATISTICAL YEARBOOK

Missouri, oil and natural gas production, reserves, exploration, development, prices, and related data, annual. OIL PRODUCING INDUSTRY IN YOUR STATE

Missouri, oil companies' market shares, by company. NATIONAL PETROLEUM NEWS FACT BOOK

Mobile (Ala.), distillates and residual fuel oils refinery and terminal prices, high, low, and average. PLATT'S OIL PRICE HANDBOOK AND OILMANAC

Mobile (Ala.), motor gasoline refinery and terminal prices, regular and premium. PLATT'S OIL PRICE HANDBOOK AND OILMANAC

Mobile loading machines at underground bituminous coal mines. BITUMINOUS COAL DATA

Mobil Oil Corporation, crude oil prices (posted). PLATT'S OIL PRICE HANDBOOK AND OILMANAC

Montana, crude production; new wells drilled and daily average production per well; natural gas and liquids production; and reserves of crude, natural gas, and liquids. TWENTIETH CENTURY PETROLEUM STATISTICS

Montana, installed generating capacity, electricity generation by type of prime mover, and electric energy sold by type of customer. EDISON ELECTRIC INSTITUTE STATISTICAL YEARBOOK

Montana, oil and natural gas production, reserves, exploration, development, prices, and related data, annual. OIL PRODUCING INDUSTRY IN YOUR STATE

Montana, oil companies' market shares, by company. NATIONAL PETROLEUM NEWS FACT BOOK

Montana, total oil and gas wells, dry holes, development, and exploratory wells drilled, by quarter. QUARTERLY REVIEW OF DRILLING STATISTICS

Moody's average yields on utility bonds and stocks, by rating and stock quality groupings. EDISON ELECTRIC INSTITUTE STATISTICAL YEARBOOK

Moody's composite average of yields on newly issued public utility bonds, by rating, monthly and annual. MOODY'S PUBLIC UTILITY MANUAL

Moody's new utility preferred stock averages, list of stocks, used in computation. MOODY'S PUBLIC UTILITY MANUAL

Moody's public utility averages, list of stocks used in computation. MOODY'S PUBLIC UTILITY MANUAL

Moody's utility bond yields by rating groups, monthly and annual. MOODY'S PUBLIC UTILITY MANUAL

Moody's utility common stocks, end-of-the-month averages (market price, dividend rate, earnings, and yield). EDISON ELECTRIC INSTITUTE STATISTICAL YEARBOOK

Moody's utility common stocks, weighted end-of-the-month averages, monthly and annual. MOODY'S PUBLIC UTILITY MANUAL

Moody's weighted averages of market price, dividend, and yield per share of transmission companies, monthly and annual. MOODY'S PUBLIC UTILITY MANUAL

Moody's weighted averages of yields on newly issued domestic utility bonds and preferred stocks, annual. MOODY'S PUBLIC UTILITY MANUAL

Morocco, crude production, refining capacity and demand for refined products. TWENTIETH CENTURY PETROLEUM STATISTICS

Motor carriers, energy freight traffic and gross freight revenue. ENERGY STATISTICS

Motorcycles, registrations by state. NATIONAL PETROLEUM FACT BOOK

Motor fuel, domestic demand, annual. TWENTIETH CENTURY PETROLEUM STATISTICS

Motor fuel, domestic demand, production, exports, imports, supply, and percent refinery yield. TWENTIETH CENTURY PETROLEUM STATISTICS

Motor fuel, highway use by passenger vehicles, buses, and cargo vehicles. ENERGY STATISTICS

Motor gasoline. See also Gasoline

Motor gasoline, average regional retail selling prices and dealer margins at full-service retail outlets. MONTHLY ENERGY REVIEW

Motor gasoline, average selling and purchase prices and dealer margins at retail outlets. MONTHLY ENERGY REVIEW

Motor gasoline, average selling prices and margins at major and independent retail dealers. MONTHLY ENERGY REVIEW

Motor gasoline, average wholesale prices, historical data. PLATT'S OIL PRICE HANDBOOK AND OILMANAC

Motor gasoline, domestic consumption. MONTHLY ENERGY REVIEW

Motor gasoline, domestic consumption by state, annual. BASIC PETROLEUM DATA BOOK

Motor gasoline, domestic consumption estimates by state. OIL PRODUCING INDUSTRY IN YOUR STATE

Motor gasoline, domestic demand and supply, daily averages. ANNUAL STATISTICAL REVIEW

Motor gasoline, production, imports, exports, and total internal consumption in OECD member countries, annual. STATISTICS OF ENERGY

Motor gasoline, production, imports, exports, consumption, and supply in OECD countries, annual and quarterly. QUARTERLY OIL STATISTICS

Motor gasoline, production, trade, and total and per capita apparent consumption, by country, annual. WORLD ENERGY SUPPLIES

Motor gasoline, production in the free world, percentages. INTERNATIONAL PETROLEUM

Motor gasoline, refinery and terminal prices, high, low, and average, regular and premium. PLATT'S OIL PRICE HANDBOOK AND OILMANAC

Motor gasoline, wholesale prices per gallon, annual. NATIONAL PETROLEUM NEWS FACT BOOK

Motor gasoline, world cargo prices. PLATT'S OIL PRICE HANDBOOK AND OILMANAC

Motor oil brand names by company. NATIONAL PETROLEUM NEWS FACT BOOK

Motor oil ratios for service stations. NATIONAL PETROLEUM NEWS FACT BOOK

Motor spirit, conversion factors. BASIC PETROLEUM DATA BOOK

Motor trucks, domestic, total registrations and vehicle miles, average miles per driven travel and average miles per gallon, annual. BASIC PETROLEUM DATA BOOK

Motor vehicle registrations by state. NATIONAL PETROLEUM NEWS FACT BOOK

Motor vehicle registrations in the United States. TWENTIETH CENTURY PETRO-
LEUM STATISTICS

Motor vehicles, domestic, total and average consumption of motor fuel, annual.
ENERGY STATISTICS

Motor vehicles, domestic, total and average consumption of motor fuel, by
type of vehicle, annual. BASIC PETROLEUM DATA BOOK

Mt. Belview (Tex.), propane refinery and terminal prices, high, low, and
average. PLATT'S OIL PRICE HANDBOOK AND OILMANAC

Municipal utilities. See Electric utilities, publicly owned

Naptha, domestic production, annual. STANDARD AND POOR'S INDUSTRY
SURVEYS

Naphtha, European bulk prices. PLATT'S OIL PRICE HANDBOOK AND OIL-
MANAC

Naphtha, forecast of selected statistical data and annual growth rates for the
United States. PREDICASTS

Naphtha, forecast of selected statistical data and annual growth rates for the
world, by region and country. WORLD CASTS

Naphtha, production, imports, exports, consumption, and stocks in OECD
countries, annual and quarterly. QUARTERLY OIL STATISTICS

Naphtha, world production by country, annual. WORLD ENERGY SUPPLIES

Natural gas, annual industrial consumption in the United States, by type of
industry. GAS FACTS

Natural gas, as a percentage of total U.S. energy resources, annual. STAN-
DARD AND POOR'S TRADE AND SECURITIES STATISTICS

Natural gas, associated, dissolved, and proved reserves in United States,
annual. ANNUAL STATISTICAL REVIEW

Natural gas, capital and exploration expenditures and gross and net fixed
assets, by world petroleum industry, by select countries. CAPITAL INVEST-
MENTS IN THE PETROLEUM INDUSTRY

Natural gas, changes in estimated proved recoverable reserves, by state. GAS
FACTS

Natural gas, coke over, gas works and refinery gas, production, trade and
total and per capita apparent consumption, by country, annual. WORLD
ENERGY SUPPLIES

Natural gas, contribution to freight revenue of U.S. railroads, annual.
ENERGY STATISTICS

Natural gas, declining annual supply in the United States.   BASIC PETROLEUM
DATA BOOK

Natural gas, demand by sector, fuel versus nonfuel use, annual.   BASIC
PETROLEUM DATA BOOK

Natural gas, demand for input in the U.S. industrial, household, commercial,
and transportation sectors, annual.   BASIC PETROLEUM DATA BOOK

Natural gas, dollar value at the wellhead, annual.   BASIC PETROLEUM
DATA BOOK

Natural gas, domestic and foreign production and consumption, world reserves,
and value at wells and at points of consumption.   TWENTIETH CENTURY
PETROLEUM STATISTICS

Natural gas, domestic marketed production and consumption, by state, annual.
BASIC PETROLEUM DATA BOOK

Natural gas, domestic production and reserves, annual.   STANDARD AND
POOR'S INDUSTRY SURVEYS

Natural gas, domestic production by state, annual.   RESERVES OF CRUDE OIL,
NATURAL GAS LIQUID AND NATURAL GAS

Natural gas, domestic recoverable reserves, by state.   BITUMINOUS COAL
DATA

Natural gas, domestic supply and demand.   ENERGY STATISTICS

Natural gas, dry, domestic consumption, annual.   COAL FACTS

Natural gas, dry, domestic consumption by household, commercial industrial,
transportation, and electricity generation users, annual.   GAS FACTS

Natural gas, estimated domestic daily production capacity during the heating
season.   RESERVES OF CRUDE OIL, NATURAL GAS LIQUID AND NATURAL
GAS

Natural gas, estimated domestic reserves by state.   OIL AND GAS CHEMICAL
SERVICE

Natural gas, estimated gross production by state.   GAS FACTS

Natural gas, estimated net production by state.   GAS FACTS

Natural gas, estimated proved reserves in the United States and other selected
areas, annual.   BASIC PETROLEUM DATA BOOK

Natural gas, estimated reserves, annual. GAS FACTS; MOODY'S PUBLIC UTILITY MANUAL

Natural gas, estimated reserves by world region and country. INTERNATIONAL OIL DEVELOPMENTS; OIL AND GAS CHEMICAL SERVICE

Natural gas, estimated supply and marketed production, value at wells, and cost to user, annual. STANDARD AND POOR'S INDUSTRY SURVEYS

Natural gas, estimated total proved reserves in the United States. RESERVES OF CRUDE OIL, NATURAL GAS LIQUID AND NATURAL GAS

Natural gas, estimated ultimate recovery in the United States, by state, by reservoir lithology, by type of entrapment, by geologic age of the reservoir, and by year of discovery. RESERVES OF CRUDE OIL, NATURAL GAS LIQUID AND NATURAL GAS

Natural gas, estimated world production, on- and offshore, by world area and by the United States, annual. BASIC PETROLEUM DATA BOOK

Natural gas, estimated world reserves by twenty leading nations. BASIC PETROLEUM DATA BOOK

Natural gas, forecast of selected statistical data and annual growth rate for the United States. PREDICASTS

Natural gas, forecast of selected statistical data and annual growth rates for the world, by region and country. WORLD CASTS

Natural gas, free-world demand, annual totals. STANDARD AND POOR'S INDUSTRY SURVEYS

Natural gas, future supplies and estimated domestic reserves, by company. OIL AND GAS CHEMICAL SERVICE

Natural gas, historical estimates of proved reserves, nationally and by state. RESERVES OF CRUDE OIL, NATURAL GAS LIQUID AND NATURAL GAS

Natural gas, imports into the United States. GAS FACTS

Natural gas, imports into the United States by source, annual. BASIC PETROLEUM DATA BOOK

Natural gas, marketed production, exports, imports, consumption, and value at wellhead, annual. BASIC PETROLEUM DATA BOOK

Natural gas, marketed production, field use, loss, waste, and loss in transmission, annual. MOODY'S PUBLIC UTILITY MANUAL

Natural gas, marketed production and average wellhead price. OIL AND GAS CHEMICAL SERVICE

Natural gas, marketed production and average wellhead price, annual. MOODY'S PUBLIC UTILITY MANUAL

Natural gas, marketed production and interstate shipments, receipts, and deliveries, by state. GAS FACTS

Natural gas, marketed production and value at wells in the United States, annual totals. STANDARD AND POOR'S TRADE AND SECURITIES STATISTICS

Natural gas, marketed production, by state. GAS FACTS

Natural gas, monthly domestic consumption, marketed production, imports and domestic producer sales to major interstate pipelines. MONTHLY ENERGY REVIEW

Natural gas, monthly underground storage. MONTHLY ENERGY REVIEW

Natural gas, nonassociated and associated-dissolved, estimated total proved reserves in the United States. RESERVE OF CRUDE OIL, NATURAL GAS LIQUID AND NATURAL GAS

Natural gas, nonassociated proved, domestic reserves, annual. ANNUAL STATISTICAL REVIEW

Natural gas, number of underground storage pools and ultimate capacity, annual. MOODY'S PUBLIC UTILITY MANUAL

Natural gas, price at wellhead, annual averages and marketed production by state. GAS FACTS

Natural gas, price at wellhead in the United States, annual average in current and constant dollars. BASIC PETROLEUM DATA BOOK

Natural gas, prices at wellhead, annual averages. STANDARD AND POOR'S INDUSTRY SURVEYS

Natural gas, production at wells, marketed production, field use, loss in transmission, and net marketed production. ENERGY STATISTICS

Natural gas, production by gas, oil wells, and repressuring, annual. MOODY'S PUBLIC UTILITY MANUAL

Natural gas, production by selected countries, annual. INTERNATIONAL ECONOMIC REPORT OF THE PRESIDENT

Natural gas, production by twenty leading nations. BASIC PETROLEUM DATA BOOK

Natural gas, production imports, exports, consumption, and supply in OECD countries, annual and quarterly. QUARTERLY OIL STATISTICS

Natural gas, production in the United States, annual totals. STANDARD AND POOR'S TRADE AND SECURITIES STATISTICS

Natural gas, production in the United States and as a percent of total energy production, annual. BASIC PETROLEUM DATA BOOK

Natural gas, proved annual reserves, total. ANNUAL STATISTICAL REVIEW

Natural gas, proved annual reserves in the United States. ANNUAL STATISTICAL REVIEW

Natural gas, proved reserves in Canada, annual. GAS FACTS

Natural gas, proved reserves in the United States, by state. ENERGY STATISTICS

Natural gas, residential service, number of customers, average annual use, average annual bill, average revenue per million BTUs, and saturation, annual. STANDARD AND POOR'S INDUSTRY SURVEYS

Natural gas, sales of producers to interstate transmission companies, by state. GAS FACTS

Natural gas, value at the wellhead, by state, annual. BASIC PETROLEUM DATA BOOK

Natural gas, value at the wellhead in the United States, total value and average cents per MCF. ANNUAL STATISTICAL REVIEW

Natural gas, value at wells, by state. OIL PRODUCING INDUSTRY IN YOUR STATE

Natural gas, world, regional and by country consumption, per capita and total. WORLD ENERGY SUPPLIES

Natural gas, world, regional, and by country production, annual totals. WORLD ENERGY SUPPLIES

Natural gas, world consumption through 1990 and as a percent of total energy consumption. BASIC PETROLEUM DATA BOOK

Natural gas, world marketed production by area, annual. BASIC PETROLEUM DATA BOOK

Natural gas and natural gas liquids, total value produced by state. OIL PRODUCING INDUSTRY IN YOUR STATE

Natural gas consumption by customers. BITUMINOUS COAL DATA

Natural gas consumption by the world by country or area. BP STATISTICAL REVIEW OF THE WORLD OIL INDUSTRY

Natural gas consumption in the United States. MONTHLY ENERGY REVIEW

Natural gas consumption in the United States, annual. TWENTIETH CENTURY PETROLEUM STATISTICS

Natural gas consumption in the United States, by type of customer. TWEN-TIETH CENTURY PETROLEUM STATISTICS

Natural gas consumption in the United States per customer, by class of service, annual. GAS FACTS

Natural gas consumption in the United States through the year 2000, by con-suming sector and as a percent of total energy input. BASIC PETROLEUM DATA BOOK

Natural gas consumption per capita by resident population in the United States, annual. GAS FACTS

Natural gas distribution companies, gas sales, number of customers, and sales ranking, by company. MOODY'S PUBLIC UTILITY MANUAL; OIL AND GAS CHEMICAL SERVICE

Natural gas distribution companies, market price, dividend rate and yield per share, Moody's weighted averages, monthly and annual. MOODY'S PUBLIC UTILITY MANUAL

Natural gas distribution companies, ranking by revenues. OIL AND GAS CHEMICAL SERVICE

Natural gas distributors, capitalization, income statement, balance sheet; and exploration and production data, by company. MOODY'S PUBLIC UTILITY MANUAL; OIL AND GAS CHEMICAL SERVICE

Natural gas distributors, composite financial and related data, annual. MOODY'S PUBLIC UTILITY MANUAL; OIL AND GAS CHEMICAL SERVICE; STANDARD AND POOR'S INDUSTRY SURVEYS

Natural gas exports from the United States. GAS FACTS

Natural gas in Canada, estimated proved remaining marketable reserves, by province. GAS FACTS

Natural gas industry, list of companies used in computation of Moody's weighted averages. MOODY'S PUBLIC UTILITY MANUAL

Natural gas industry stocks, market price, dividend rate, and yields per share, Moody's weighted averages, monthly and annual. MOODY'S PUBLIC UTILITY MANUAL

Natural gas in the United States, gross withdrawals and disposition, by state, daily averages. ANNUAL STATISTICAL REVIEW

Natural gas in the United States, production and disposition. GAS FACTS

Natural gas in the United States, supply and demand, daily averages. AN-NUAL STATISTICAL REVIEW

Natural gas liquids, changes in estimated proved recoverable reserves, by state. GAS FACTS

Natural gas liquids, consumption in the United States, annual. COAL FACTS

Natural gas liquids, domestic production, demand, and stocks, annual. STAN-DARD AND POOR'S TRADE AND SECURITIES STATISTICS

Natural gas liquids, domestic recoverable reserves, by state. BITUMINOUS COAL DATA

Natural gas liquids, estimated domestic daily productive capacity during the heating season. RESERVES OF CRUDE OIL, NATURAL GAS LIQUID AND NATURAL GAS

Natural gas liquids, estimated domestic reserves, by state. OIL AND GAS CHEMICAL SERVICE

Natural gas liquids, estimated reserves, annual. GAS FACTS

Natural gas liquids, estimated total proved reserves in the United States. RESERVES OF CRUDE OIL, NATURAL GAS LIQUID AND NATURAL GAS

Natural gas liquids, estimates of proved reserves in the United States, annual. ENERGY STATISTICS

Natural gas liquids, Federal Reserve Board indexes of industrial production, monthly and annual averages. STANDARD AND POOR'S TRADE AND SECURITIES STATISTICS

Natural gas liquids, forecast of selected statistical data and annual growth rates for the United States. PREDICASTS

Natural gas liquids, forecast of selected statistical data and annual growth rates for the world, by region and country. WORLD CASTS

Natural gas liquids, historical estimates of proved reserves, nationally and by state. RESERVES OF CRUDE OIL, NATURAL GAS LIQUID AND NATURAL GAS

Natural gas liquids, monthly domestic demand, production, imports, and stocks, MONTHLY ENERGY REVIEW

Natural gas liguids, production, imports, exports, consumption, and stocks in OECD countries, annual and quarterly. QUARTERLY OIL STATISTICS

Natural gas liquids, production in the United States, annual, historical data. OIL PRODUCING INDUSTRY IN YOUR STATE

Natural gas liquids, proved annual reserves in the United States. ANNUAL STATISTICAL REVIEW

Natural gas liquids, total U.S. supply and demand, annual. BASIC PETRO-LEUM DATA BOOK

Natural gas liquids, value produced, by state. OIL PRODUCING INDUSTRY IN YOUR STATE

Natural gas liquids, world production by country and region in barrels per day. INTERNATIONAL OIL DEVELOPMENTS

Natural gas liquids and liquefied refinery gases, total domestic supply and demand, daily averages. ANNUAL STATISTICAL REVIEW

Natural gasoline, production, trade, and total and per capita apparent consumption, by country, annual. WORLD ENERGY SUPPLIES

Natural gasoline, production in the United States, annual totals. STANDARD AND POOR'S TRADE AND SECURITIES STATISTICS

Natural gasoline, total blended at stills, domestic production, and stocks and value at plants. TWENTIETH CENTURY PETROLEUM STATISTICS

Natural gas pipelines and utility mains, by state and region. ENERGY STATISTICS

Natural gas producers, capitalization, income statement, balance sheet and exploration and production data, by company. MOODY'S PUBLIC UTILITY MANUAL; OIL AND GAS CHEMICAL SERVICE

Natural gas producers, sales revenues by sales to interstate transmission companies, by state. GAS FACTS

Natural gas sources, domestic and imported, of the United States, Japan, European community, United Kingdom, France, West Germany, and Italy. INTERNATIONAL ECONOMIC REPORT OF THE PRESIDENT

Natural gas transportation companies, ranking by revenues. OIL AND GAS CHEMICAL SERVICE

Natural gas treated for natural gasoline and allied products, and quantities and value of products recovered. GAS FACTS

Natural gas used as pipeline fuel in the United States, annual. GAS FACTS

Natural gas wells, number under operation, by state. ANNUAL STATISTICAL REVIEW

Natural gas (wet), production and consumption from water and nuclear power, annual. BITUMINOUS COAL DATA

Natural gas (wet), production in the United States, annual. COAL FACTS

Natural gas (wet and unprocessed), production in the United States, annual. GAS FACTS

Nebraska, crude production; new wells drilled and daily average production per well; natural gas and liquids production; and reserves of crude, natural gas, and liquids. TWENTIETH CENTURY PETROLEUM STATISTICS

Nebraska, gasoline and fuel oil prices. OIL DAILY

Nebraska, installed generating capacity, electricity generation by type of customer, and electric energy sold by type of customer. EDISON ELECTRIC INSTITUTE STATISTICAL YEARBOOK

Nebraska, oil and natural gas production, reserves, exploration, development, prices, and related data, annual. OIL PRODUCING INDUSTRY IN YOUR STATE

Nebraska, oil companies' market shares, by company. NATIONAL PETROLEUM NEWS FACT BOOK

Nebraska, total oil and gas wells, dry holes, development, and exploratory wells drilled, by quarter. QUARTERLY REVIEW OF DRILLING STATISTICS

Netherlands, crude production, total and as a percent of world production, reserves of crude oil and natural gas, refining capacity, refined products demand, and tank ship fleet. TWENTIETH CENTURY PETROLEUM STATISTICS

Netherlands, estimated imports of crude and refined products from Arab and non-Arab countries by original crude source. INTERNATIONAL OIL DEVELOP-MENTS

Netherlands, production, consumption, exports and imports by direction, of crude oil and crude oil products by type, and of natural gas and liquids. QUARTERLY OIL STATISTICS

Netherlands, production, imports, exports, and consumption by function and user, of energy sources by type. STATISTICS OF ENERGY

Neutral oils (solvent refined), refinery and terminal prices, select regions. PLATT'S OIL PRICE HANDBOOK AND OILMANAC

Nevada, crude production, new wells drilled and daily average production at wells. TWENTIETH CENTURY PETROLEUM STATISTICS

Nevada, installed generating capacity, electricity generation by type of prime mover, and electric energy sold by type of customer. EDISON ELECTRIC INSTITUTE STATISTICAL YEARBOOK

Nevada, oil and natural gas production, reserves, exploration, development, prices, and related data, annual. OIL PRODUCING INDUSTRY IN YOUR STATE

Nevada, oil companies' market shares, by company. NATIONAL PETROLEUM NEWS FACT BOOK

Nevada, total oil and gas wells, dry holes, development, and exploratory wells drilled by quarter. QUARTERLY REVIEW OF DRILLING STATISTICS

New field discoveries, oil and gas, total exploratory wells drilled, and new oil reserves added, annual. STANDARD AND POOR'S INDUSTRY SURVEYS

New Hampshire, installed generating capacity, electricity generation by type of prime mover, and electric energy sold by type of customer. EDISON ELECTRIC INSTITUTE STATISTICAL YEARBOOK

New Hampshire, oil companies' market shares, by company. NATIONAL PETROLEUM NEWS FACT BOOK

New Haven (Conn.), distillates and residual fuel oils cargo prices, high, low, and average. PLATT'S OIL PRICE HANDBOOK AND OILMANAC

New Haven (Conn.), kerosene refinery and terminal prices, high, low, and average. PLATT'S OIL PRICE HANDBOOK AND OILMANAC

New Jersey, installed generating capacity, electricity generation by type of prime mover, and electric energy sold by type of customer. EDISON ELECTRIC INSTITUTE STATISTICAL YEARBOOK

New Jersey, oil companies' market shares, by company. NATIONAL PETRO-LEUM NEWS FACT BOOK

New Mexico, crude production; new wells drilled and daily average production at wells; natural gas and liquids production; and reserves of crude, natural gas, and liquids. TWENTIETH CENTURY PETROLEUM STATISTICS

New Mexico, installed generating capacity, electricity generation by type of prime mover, and electric energy sold by type of customer. EDISON ELEC-TRIC INSTITUTE STATISTICAL YEARBOOK

New Mexico, oil and natural gas production, reserves, exploration, develop-ment, prices, and related data, annual. OIL PRODUCING INDUSTRY IN YOUR STATE

New Mexico, oil companies' market shares, by company. NATIONAL PETRO-LEUM NEWS FACT BOOK

New Mexico, total oil and gas wells, dry holes, development, and exploratory wells drilled, by quarter. QUARTERLY REVIEW OF DRILLING STATISTICS

New Orleans (La.), kerosene and marine diesel refinery and terminal prices, high, and low, and average. PLATT'S OIL PRICE HANDBOOK AND OIL-MANAC

New Orleans, (La.), motor gasoline refinery and terminal prices, regular and premium. PLATT'S OIL PRICE HANDBOOK AND OILMANAC

New York, crude production, new wells drilled and daily average production per well, natural gas and gasoline production, and reserves of natural gas and crude. TWENTIETH CENTURY PETROLEUM STATISTICS

New York, fuel oil prices. OIL DAILY

New York, gasoline prices, premium, unleaded, and regular. OIL DAILY

New York, installed generating capacity, electricity generation by type of prime mover, and electric energy sold by type of customer. EDISON ELECTRIC INSTITUTE STATISTICAL YEARBOOK

New York, oil and natural gas production, reserves, exploration, development, prices, and related data, annual. OIL PRODUCING INDUSTRY IN YOUR STATE

New York, oil companies' market shares, by company. NATIONAL PETROLEUM NEWS FACT BOOK

New York, propane refinery and terminal prices, high, low, and average. PLATT'S OIL PRICE HANDBOOK AND OILMANAC

New York, total oil and gas wells, dry holes, development, and exploratory wells drilled, by quarter. QUARTERLY REVIEW OF DRILLING STATISTICS

New York City, motor gasoline refinery and terminal prices, regular and premium. PLATT'S OIL PRICE HANDBOOK AND OILMANAC

New York Harbor, distillate and residual fuel oil, cargo prices, high, low, and average. PLATT'S OIL PRICE HANDBOOK AND OILMANAC

New Zealand, production, consumption, exports and imports by direction, of crude oil and crude oil products by type, and of natural gas and liquids. QUARTERLY REVIEW OF DRILLING STATISTICS

New Zealand, production, imports, exports, and consumption, by function and user, of energy sources by type. STATISTICS OF ENERGY

Nigeria, crude oil prices by oil field. PLATT'S OIL PRICE HANDBOOK AND OILMANAC

Nigeria, crude oil production in barrels per day. INTERNATIONAL OIL DE-
VELOPMENTS

Nigeria, crude oil production in barrels per day and as a percent of total
production. INTERNATIONAL ECONOMIC REPORT OF THE PRESIDENT

Nigeria, crude production, total and as a percent of world production, exports
to the United States, reserves of crude and natural gas, and demand for refined products. TWENTIETH CENTURY PETROLEUM STATISTICS

Nigeria, imports from and exports to the United States and other developed
countries, total value. INTERNATIONAL OIL DEVELOPMENTS

Nigeria, oil reserves and production, exports of crude oil and petroleum to
the United States, annual. BASIC PETROLEUM DATA BOOK

Nonenergy petroleum products, world production by country, annual. WORLD
ENERGY SUPPLIES

Norfolk (Va.), motor gasoline refinery and terminal prices, regular and premium.
PLATT'S OIL PRICE HANDBOOK AND OILMANAC

Norfolk (Va.), refinery and terminal prices of distillate and residual fuel oils.
PLATT'S OIL PRICE HANDBOOK AND OILMANAC

North America, crude oil production in barrels per day and as a percent of
total production. INTERNATIONAL ECONOMIC REPORT OF THE PRESIDENT

North America, crude production, total and as a percent of world production,
demand production ratio, reserves of natural gas and crude, refining capacity,
demand for refined products, and tank ship fleet. TWENTIETH CENTURY
PETROLEUM STATISTICS

North America, export of solid fuel to other countries, by country, annual.
WORLD ENERGY SUPPLIES

North America, exports of crude petroleum to countries, by country, annual.
WORLD ENERGY SUPPLIES

North American continent, production and use of energy sources by type of
energy and by country. STATISTICS OF ENERGY

North Carolina, footage of new wells drilled. TWENTIETH CENTURY PETRO-
LEUM STATISTICS

North Carolina, installed generating capacity, electricity generation by type of
prime mover, and electric energy sold by type of prime mover. EDISON
ELECTRIC INSTITUTE STATISTICAL YEARBOOK

North Carolina, oil companies' market news, by company. NATIONAL PETRO-
LEUM NEWS FACT BOOK

North Carolina, total oil and gas wells, dry holes, development, and exploratory wells drilled, by quarter. QUARTERLY REVIEW OF DRILLING STATISTICS

North Dakota, crude production; new wells drilled and average daily production per well; natural gas and liquids production; and reserves of crude, natural gas, and liquids. TWENTIETH CENTURY PETROLEUM STATISTICS

North Dakota, installed generating capacity, electricity generation by type of prime mover, and electric energy sold by type of prime mover. EDISON ELECTRIC INSTITUTE STATISTICAL YEARBOOK

North Dakota, oil and natural gas production, reserves, exploration, and development, prices, and related data. OIL PRODUCING INDUSTRY IN YOUR STATE

North Dakota, oil companies' market shares, by company. NATIONAL PETROLEUM NEWS FACT BOOK

North Dakota, total oil and gas wells, dry holes, development, and exploratory wells drilled, by quarter. QUARTERLY REVIEW OF DRILLING STATISTICS

North Sea Oil, estimated reserves and production, by field and type of operator. STANDARD AND POOR'S INDUSTRY SURVEYS

North Texas, gasoline and fuel oil prices. OIL DAILY

Norway, crude oil production in barrels per day. INTERNATIONAL OIL DEVELOPMENTS

Norway, crude oil production in barrels per day and as a percent of total production. INTERNATIONAL ECONOMIC REPORT OF THE PRESIDENT

Norway, crude production, total and as a percent of world production, reserves of crude and natural gas, refining capacity, refined products demand, and tank ship fleet. TWENTIETH CENTURY PETROLEUM STATISTICS

Norway, natural gas liquids production in barrels per day. INTERNATIONAL OIL DEVELOPMENTS

Norway, production, consumption, and exports and imports by direction, of crude oil and products by type, and of natural gas and liquids. QUARTERLY OIL STATISTICS

Norway, production, imports, exports, and consumption by function and user, of energy sources by type. STATISTICS OF ENERGY

Nuclear electric power, forecast of selected statistical data and annual growth rates for the United States. PREDICASTS

Nuclear electric power, forecast of selected statistical data and annual growth rates for the world, by region and country. WORLD CASTS

Nuclear electric power, total free-world demand, annual. STANDARD AND POOR'S INDUSTRY SURVEYS

Nuclear electronics industry, value of factory sales, annual. STANDARD AND POOR'S INDUSTRY SURVEYS

Nuclear energy, installed generating capacity of selected countries, annual. INTERNATIONAL ECONOMIC REPORT OF THE PRESIDENT

Nuclear-fuel-generated electricity, total amount, annual and by state. EDISON ELECTRIC INSTITUTE STATISTICAL YEARBOOK

Nuclear generating plants, in service, under construction or announced, station, plant rating, and service data. MOODY'S PUBLIC UTILITY MANUAL

Nuclear power, annual production in the United States. GAS FACTS

Nuclear power, annual U.S. consumption for electricity generation. GAS FACTS

Nuclear power, consumption by the world, by country or area. BP STATISTICAL REVIEW OF THE WORLD OIL INDUSTRY

Nuclear power, consumption in the United States through the year 2000, by consuming sector and as a percent of total energy input. BASIC PETROLEUM DATA BOOK

Nuclear power, electricity production in the United States. TWENTIETH CENTURY PETROLEUM STATISTICS

Nuclear power, growth estimate, distribution of reactor types, and reactor characteristics. OECD URANIUM RESOURCES, PRODUCTION AND DEMAND

Nuclear power, plant operations, status, uranium enrichment, power generation, and fuel cycle. MONTHLY ENERGY REVIEW

Nuclear power, production in the United States and as a percent of total energy production, annual. BASIC PETROLEUM DATA BOOK

Nuclear power, production in United States, by state. COAL FACTS

Nuclear power, projects in operation, under construction, or under design, year the contract was awarded, supplier, and type of reactor, expected initial year of operation, and operating capability. STEAM ELECTRIC PLANT FACTORS

Nuclear power, world consumption through 1990 and as a percent of total energy consumption. BASIC PETROLEUM DATA BOOK

Nuclear reactors. See Reactors (nuclear)

Nuclear steam-electric plants, reported new capability addition by type of ownership. MOODY'S PUBLIC UTILITY MANUAL

Nuclear steam-electric utilities, electricity generation by type of ownership, annual. EDISON ELECTRIC INSTITUTE STATISTICAL YEARBOOK

Nuclear steam-electric utilities, installed generating capacity by ownership, annual. EDISON ELECTRIC INSTITUTE STATISTICAL YEARBOOK

Nuclear steam-generating plants, kilowatts generated and as a percent of total production, annual. STANDARD AND POOR'S INDUSTRY SURVEYS

OAPEC countries, crude oil production and capacities. INTERNATIONAL OIL DEVELOPMENTS

OECD countries, production, consumption, and exports and imports by direction, of crude oil and crude oil products, by type, and of natural gas and liquids. QUARTERLY OIL STATISTICS

OECD countries, production and uses of energy sources by type of source, by member country. STATISTICS OF ENERGY

Offshore drilling companies, domestic, composite financial and related data, annual. STANDARD AND POOR'S INDUSTRY SURVEYS

Offshore drilling operations and expenditures by type of well. GAS FACTS

Offshore oil wells drilled in the United States, by state. BASIC PETROLEUM DATA BOOK

Offshore wells, drilling and equipping costs in the United States. TWENTIETH CENTURY PETROLEUM STATISTICS

Ohio, crude production; new wells drilled and daily average production per well; gasoline, natural gas, and liquids production; and reserves of crude, natural gas, and liquids. TWENTIETH CENTURY PETROLEUM STATISTICS

Ohio, installed generating capacity, electricity generation by type of prime mover, and electric energy sold by type of customer. EDISON ELECTRIC INSTITUTE STATISTICAL YEARBOOK

Ohio, oil and natural gas production, reserves, exploration, development, prices, and related data, annual. OIL PRODUCING INDUSTRY IN YOUR STATE

Ohio, oil companies' market shares, by company. NATIONAL PETROLEUM NEWS FACT BOOK

Ohio, total oil and gas wells, dry holes, development, and exploratory wells drilled by quarter. QUARTERLY REVIEW OF DRILLING STATISTICS

Oil, as a percentage of total U.S. energy sources, annual. STANDARD AND POOR'S TRADE AND SECURITIES STATISTICS

Oil, calorific equivalent in electricity. BP STATISTICAL REVIEW OF THE WORLD OIL INDUSTRY

Oil, calorific equivalent in heat units. BP STATISTICAL REVIEW OF THE WORLD OIL INDUSTRY

Oil, calorific equivalent in natural gas. BP STATISTICAL REVIEW OF THE WORLD OIL INDUSTRY

Oil, calorific equivalent in solid fuels. BP STATISTICAL REVIEW OF THE WORLD OIL INDUSTRY

Oil, calorific equivalent in town gas. BP STATISTICAL REVIEW OF THE WORLD OIL INDUSTRY

Oil, first year of production in each state. OIL PRODUCING INDUSTRY IN YOUR STATE

Oil and gas (acres), area proved productive in each state. OIL PRODUCING INDUSTRY IN YOUR STATE

Oil and gas drilling industry, Federal Reserve Board indexes of industrial production, monthly and annual averages. STANDARD AND POOR'S TRADE AND SECURITIES STATISTICS

Oil and gas exploratory activities in the United States. MONTHLY ENERGY REVIEW

Oil and gas extraction industry, hourly wages, weekly earnings, and hours worked, monthly and annual averages. STANDARD AND POOR'S TRADE AND SECURITIES STATISTICS

Oil and gas production, estimated productive and nonproductive acreage under lease, by PAD district. OIL PRODUCING INDUSTRY IN YOUR STATE

Oil and gas wells, number of drilled wells, by PAD district. OIL PRODUCING INDUSTRY IN YOUR STATE

Oil (coal equivalent) consumption by steam-electric plant, by region. COAL FACTS

Oil companies, appriased value per common share, by company. OIL AND GAS CHEMICAL SERVICE

Oil companies, capitalization, income statement and balance sheet, and exploration and production data, by company. OIL AND GAS CHEMICAL SERVICE

Oil companies, comparative rate of return on investment, by company. OIL AND GAS CHEMICAL SERVICE

Oil companies, composite financial statistics, annual. STANDARD AND POOR'S INDUSTRY SURVEYS

Oil companies, distribution channels for gasoline and products, description and total number, by company. NATIONAL PETROLEUM NEWS FACT BOOK

Oil companies, domestic, gross sales, operating revenue, net income, total assets, and net income as a percent of stockholders equity. NATIONAL PETRO-LEUM NEWS FACT BOOK

Oil companies, domestic, net foreign production of crude and refinery runs by world area. OIL AND GAS CHEMICAL SERVICE

Oil companies, domestic, net liquid production, refinery runs, and product sales, by company. OIL AND GAS CHEMICAL SERVICE

Oil companies, domestic, oil products and crude imports, net crude production, refinery crude runs, and domestic refined petroleum product sales. NATIONAL PETROLEUM NEWS FACT BOOK

Oil companies, domestic, ranking by assets, net income, and net sales. NATIONAL PETROLEUM NEWS FACT BOOK

Oil companies, domestic, stockholders equity, and profitability ratios, by company. OIL AND GAS CHEMICAL SERVICE

Oil companies, domestic, total number of branded retail outlets and service stations, by company. NATIONAL PETROLEUM NEWS FACT BOOK

Oil companies, domestic market shares of motor gasoline sales, by company and state. NATIONAL PETROLEUM NEWS FACT BOOK

Oil company stocks average daily closing price and annual price range. OIL DAILY

Oil consumption, domestic, by passenger cars, annual averages. NATIONAL PETROLEUM NEWS FACT BOOK

Oil consumption, domestic, for electricity generation by individual utilities, and unit costs. STEAM ELECTRIC PLANT FACTORS

Oil consumption, in selected Western countries, annual. INTERNATIONAL ECONOMIC REPORT OF THE PRESIDENT

Oil consumption, in the United States and other OECD countries, annual. INTERNATIONAL OIL DEVELOPMENTS

Oil consumption, in the world, by country or area. BP STATISTICAL REVIEW OF THE WORLD OIL INDUSTRY

Oil consumption, per capita in the United States, annual. TWENTIETH CEN-
TURY PETROLEUM STATISTICS

Oil consumption and trade in the world, by country or area, average annual
rate. BP STATISTICAL REVIEW OF THE WORLD PETROLEUM INDUSTRY

Oil demand, domestic versus the gross national product. STANDARD AND
POOR'S INDUSTRY SURVEYS

Oil discovered. See also Oil wells, new completions

Oil discovered and produced, historical. BP STATISTICAL REVIEW OF THE
WORLD OIL INDUSTRY

Oil field hardware companies, sales to petroleum industry, by company. STAN-
DARD AND POOR'S INDUSTRY SURVEYS

Oil field machinery and tools, wholesale price index, annual. BASIC PETRO-
LEUM DATA BOOK

Oil fields in the United States, 100 largest fields, with remaining estimated
proved reserves. ENERGY STATISTICS

Oil field stocks, closing prices, weekly change, price earnings, ratio and
yield. OIL DAILY

Oil heating degree days. See Degree days

Oil imports into major consuming areas, main sources. BP STATISTICAL REVIEW
OF THE WORLD OIL INDUSTRY

Oil movements, interarea by country and region. BP STATISTICAL REVIEW OF
THE WORLD OIL INDUSTRY

Oil movements by sea. BP STATISTICAL REVIEW OF THE WORLD OIL INDUS-
TRY

Oil pipelines. See Pipelines

Oil producing wells, total number of wells in the world, by country. INTER-
NATIONAL PETROLEUM

Oil production, by country and world region, annual. ENERGY STATISTICS

Oil production, by country or area. BP STATISTICAL REVIEW OF THE WORLD
OIL INDUSTRY

Oil production, by OPEC and non-OPEC countries. BP STATISTICAL REVIEW
OF THE WORLD OIL INDUSTRY

Oil production, free-world totals. BP STATISTICAL REVIEW OF THE WORLD OIL INDUSTRY

Oil production, in the United States, area proved productive in acres, by state. OIL PRODUCING INDUSTRY IN YOUR STATE

Oil production, in the world by country or area, average annual rate. BP STATISTICAL REVIEW OF THE WORLD OIL INDUSTRY

Oil products, forecast of selected statistical data and annual growth rates for the United States. PREDICASTS

Oil products, forecast of selected statistical data and annual growth rates for the world, by region and country. WORLD CASTS

Oil refining capacities of the Eastern hemisphere. BP STATISTICAL REVIEW OF THE WORLD OIL INDUSTRY

Oil refining capacities of the Western Hemisphere, annual. BP STATISTICAL REVIEW OF THE WORLD OIL INDUSTRY

Oil refining capacities of the world, by country or area, annual. BP STATISTICAL REVIEW OF THE WORLD OIL INDUSTRY

Oil refining capacities of the world, by country or area, average annual rate. BP STATISTICAL REVIEW OF THE WORLD OIL INDUSTRY

Oil reserves, published and proved, annual, by country or area. BP STATISTICAL REVIEW OF THE WORLD OIL INDUSTRY

Oil reserves, published and proved, annual, Eastern Hemisphere, total. BP STATISTICAL REVIEW OF THE WORLD OIL INDUSTRY

Oil reserves, published and proved, annual, Western Hemisphere, total. BP STATISTICAL REVIEW OF THE WORLD OIL INDUSTRY

Oil reserves, published and proved, annual, world totals. BP STATISTICAL REVIEW OF THE WORLD OIL INDUSTRY

Oil reserves of the world, proved, by selected countries and regions, in barrels and as a percent, total. INTERNATIONAL ECONOMIC REPORT OF THE PRESIDENT

Oils, domestic demand and supply, and stocks, by PAD district, daily averages. ANNUAL STATISTICAL REVIEW

Oils, domestic demand and supply, annual. BASIC PETROLEUM DATA BOOK

Oils, domestic production, imports, new supply, disposition, exports, and demand, daily averages. ANNUAL STATISTICAL REVIEW

Oils (all types), days supply, domestic demand, per capita consumption, stocks, total demand and supply, and years' supply. TWENTIETH CENTURY PETRO-LEUM STATISTICS

Oil shale consumption in the United States through the year 2000, by consuming sector and as a percent of total energy input. BASIC PETROLEUM DATA BOOK

Oil shale deposits in the United States by state. ENERGY STATISTICS

Oil stocks in the United States and other OECD countries, end-of-the-month. INTERNATIONAL OIL DEVELOPMENTS

Oil supply and demand for the world, annual. BP STATISTICAL REVIEW OF THE WORLD OIL INDUSTRY

Oil utilization at steam-electric plants, trends in efficiency. STEAM ELECTRIC PLANT FACTORS

Oil wells. See also Wells

Oil wells, average cost per foot, by depth classes. GAS FACTS

Oil wells, crude, gas, and condensate, total number in operation, by state. OIL PRODUCING INDUSTRY IN YOUR STATE

Oil wells, deepest producing well drilled, by state. OIL PRODUCING IN-DUSTRY IN YOUR STATE

Oil wells, deepest wells drilled, by state. OIL PRODUCING INDUSTRY IN YOUR STATE

Oil wells, domestic, new completions, abandonments, production per well, and stripper wells. OIL AND GAS CHEMICAL SERVICE

Oil wells, drilling and equipping, by depth intervals, estimated cost in the United States. ANNUAL STATISTICAL REVIEW

Oil wells, drilling and equipping wells, by PAD district, estimated costs. OIL PRODUCING INDUSTRY IN YOUR STATE

Oil wells, footage drilled, annual historical data. OIL PRODUCING INDUS-TRY IN YOUR STATE

Oil wells, footage drilled and estimated drilling costs, by state. ANNUAL STATISTICAL REVIEW

Oil wells, free-world completions, by country. BASIC PETROLEUM DATA BOOK

Oil wells, new completions, by state. GAS FACTS

Oil wells, number drilled, annual. STANDARD AND POOR'S TRADE AND SECURITIES STATISTICS

Oil wells, number of abandoned wells, average life, daily average production per well by state and PAD district, number of producing wells, and costs of drilling and equipment. TWENTIETH CENTURY PETROLEUM STATISTICS

Oil wells, number of abandoned wells in the United States. TWENTIETH CENTURY PETROLEUM STATISTICS

Oil wells, number of active wells, and average production per well per day, annual. STANDARD AND POOR'S INDUSTRY SURVEYS

Oil wells, number of active wells in the United States and average daily production per well, annual. STANDARD AND POOR'S TRADE AND SECURI-TIES STATISTICS

Oil wells, number of exploratory wells drilled, by PAD district. OIL PRODUC-ING INDUSTRY IN YOUR STATE

Oil wells, number of new field wildcat wells drilled, by PAD district. OIL PRODUCING INDUSTRY IN YOUR STATE

Oil wells, number of new wells drilled, historical data, annual. OIL PRODUC-ING INDUSTRY IN YOUR STATE

Oil wells, number of producing stripper and nonstripper wells and abandoned wells. BASIC PETROLEUM DATA BOOK

Oil wells, number of producing stripper wells, by selected states. OIL PRO-DUCING INDUSTRY IN YOUR STATE

Oil wells, number of wells drilled, by type of well and average depth per well. STANDARD AND POOR'S INDUSTRY SURVEYS

Oil wells, number of wells drilled in the United States. ANNUAL STATISTICAL REVIEW; STANDARD AND POOR'S TRADE AND SECURITIES STATISTICS

Oil wells, number of wells drilled in the United States, by PAD district. OIL PRODUCING INDUSTRY IN YOUR STATE

Oil wells, number of wells under operation, by state. ANNUAL STATISTICAL REVIEW

Oil wells, offshore drilling operations and expenditures. GAS FACTS

Oil wells, onshore estimated drilling costs in the United States, annual. BASIC PETROLEUM DATA BOOK

Oil wells, percent of wells pumping or on artificial lift, by state. OIL PRO-
DUCING INDUSTRY IN YOUR STATE

Oil wells, productive, average drilling costs in the United States, annual.
ANNUAL STATISTICAL REVIEW

Oil wells, productive, number of wells, footage drilled, and cost of drilling
and equipping wells and dry holes, annual. ANNUAL STATISTICAL REVIEW

Oil wells and dry holes drilled as exploratory tests by state. GAS FACTS

Oil wells and footage drilled, by state. OIL PRODUCING INDUSTRY IN
YOUR STATE

Oil wells casing alloy and carbon, wholesale price index, annual. BASIC
PETROLEUM DATA BOOK

Oil well service companies, domestic, composite financial and related data
annual. STANDARD AND POOR'S INDUSTRY SURVEYS

Oklahoma, crude production; new wells drilled and average daily production
per well; production of gasoline, natural gas, and liquids; and reserves of
crude, natural gas and liquids. TWENTIETH CENTURY PETROLEUM STATISTICS

Oklahoma, gasoline and fuel oil prices. OIL DAILY

Oklahoma, installed generating capacity, electricity generation by type of
prime mover, and electric energy sold by type of customer. EDISON ELECTRIC
INSTITUTE STATISTICAL YEARBOOK

Oklahoma, motor gasoline refinery and terminal prices, regular and premium.
PLATT'S OIL PRICE HANDBOOK AND OILMANAC

Oklahoma, oil and natural gas production, reserves, exploration, development,
prices, and related data, annual. OIL PRODUCING INDUSTRY IN YOUR
STATE

Oklahoma, oil companies' market shares, by company. NATIONAL PETROLEUM
NEWS FACT BOOK

Oklahoma, propane refinery and terminal prices, high, low, and average.
PLATT'S OIL PRICE HANDBOOK AND OILMANAC

Oklahoma, total oil and gas wells, dry holes, development, and exploratory
wells drilled by quarter. QUARTERLY REVIEW OF DRILLING STATISTICS

Oklahoma-Kansas crude oil, monthly price and annual average. STANDARD
AND POOR'S TRADE AND SECURITIES STATISTICS

Oman, crude oil prices, by oil field. PLATT'S OIL PRICE HANDBOOK AND
OILMANAC

Oman, crude oil production, total and as a percent of world production, and reserves of crude oil and natural gas. TWENTIETH CENTURY PETROLEUM STATISTICS

Oman, crude oil production in barrels per day. INTERNATIONAL OIL DEVELOPMENTS

OPEC countries, crude oil production and capacity. INTERNATIONAL OIL DEVELOPMENTS

OPEC countries, exports of petroleum to the United States, by member country. BASIC PETROLEUM DATA BOOK

OPEC countries, imports from and exports to developed countries, by country. INTERNATIONAL OIL DEVELOPMENTS

OPEC countries, oil reserves and production, exports of crude oil and petroleum to the United States, annual. BASIC PETROLEUM DATA BOOK

Operating expenses of gas and electric utilities in the United States. FINANCIAL STATISTICS OF PUBLIC UTILITIES; MOODY'S PUBLIC UTILITY MANUAL; OIL AND GAS CHEMICAL SERVICE

Operating income of gas and electric utilities in the United States. FINANCIAL STATISTICS OF PUBLIC UTILITIES; MOODY'S PUBLIC UTILITY MANUAL

Oregon, installed generating capacity, electricity generation by type of prime mover, and electric energy sold by type of customer. EDISON ELECTRIC INSTITUTE STATISTICAL YEARBOOK

Oregon, oil companies' market shares, by company. NATIONAL PETROLEUM NEWS DATA BOOK

Outer continental shelf oil and gas lease sales, revenue, and production value. BASIC PETROLEUM DATA BOOK

Output per man per day at bituminous coal mines by type, domestic and selected foreign mines. BITUMINOUS COAL DATA

Oven coke. See Coke oven coke

Oven coke plants, annual consumption of bituminous coal. BITUMINOUS COAL DATA

Overhead electric and voltage circuit miles, by state. EDISON ELECTRIC INSTITUTE STATISTICAL YEARBOOK

PAD districts. See Petroleum Administration for Defense districts

Pakistan, crude production, total and as a percent of world production, reserves of crude and natural gas, and tank ship fleet. TWENTIETH CENTURY PETROLEUM STATISTICS

Pale neutral oils, refinery and terminal prices, select regions. PLATT'S OIL PRICE HANDBOOK AND OILMANAC

Panama, demand for refined products. TWENTIETH CENTURY PETROLEUM STATISTICS

Paraffin wax, world production by country, annual. WORLD ENERGY SUPPLIES

Paraguay, total refining capacity. TWENTIETH CENTURY PETROLEUM STATISTICS

Passenger cars, average fuel efficiency in the United States, annual. ENERGY STATISTICS

Passenger cars, domestic registrations and vehicle miles, average miles per travel and average miles per gallon, annual. BASIC PETROLEUM DATA BOOK

Passenger cars, registrations in the United States. TWENTIETH CENTURY PETROLEUM STATISTICS

Pennsylvania, crude production; new wells drilled and daily average production per well; production of gasoline, natural gas, and liquids; and reserves of crude, natural gas and liquids. TWENTIETH CENTURY PETROLEUM STATISTICS

Pennsylvania, installed generating capacity, electricity generation by type of mover, and electric energy sold by type of customer. EDISON ELECTRIC INSTITUTE STATISTICAL YEARBOOK

Pennsylvania, oil and natural gas production, reserves, exploration, development, prices, and related data, annual. OIL PRODUCING INDUSTRY IN YOUR STATE

Pennsylvania, oil companies' market share, by companies. NATIONAL PETROLEUM NEWS FACT BOOK

Pennsylvania, total oil and gas wells, dry holes, development, and exploratory wells drilled, by quarter. QUARTERLY REVIEW OF DRILLING STATISTICS

Pennzoil Company, crude oil prices (posted). PLATT'S OIL PRICE HANDBOOK AND OILMANAC

Persian Gulf, crude oil prices, by oil field. PLATT'S OIL PRICE HANDBOOK AND OILMANAC

Peru, crude production, total and as a percent of world production, reserves of crude and natural gas refining capacity, refined products demand, and tank ship fleet. TWENTIETH CENTURY PETROLEUM STATISTICS

Petroleum, crude and refined stocks on hand, end-of-the-month totals. STAN-DARD AND POOR'S TRADE AND SECURITIES STATISTICS

Petroleum, domestic consumption, demand, and supply, by product and sector, annual. ENERGY STATISTICS

Petroleum, domestic consumption by household, commercial, industrial, transportation, and electricity generation users, annual. ENERGY STATISTICS

Petroleum, domestic demand, historical data, annual. OIL PRODUCING INDUSTRY IN YOUR STATE

Petroleum, domestic demand by sector, fuel versus nonfuel use, annual. BASIC PETROLEUM DATA BOOK

Petroleum, domestic demand by type, annual. STANDARD AND POOR'S INDUSTRY SURVEYS

Petroleum, domestic demand exports, annual historical data. OIL PRODUCING INDUSTRY IN YOUR STATE

Petroleum, domestic demand for input in the U.S. industrial, household, commercial, and transportation sectors, annual. BASIC PETROLEUM DATA BOOK

Petroleum, domestic exports to selected areas, annual. BASIC PETROLEUM DATA BOOK

Petroleum, domestic imports by source, and from members of OPEC countries. BASIC PETROLEUM DATA BOOK

Petroleum, domestic imports from the Eastern Hemisphere by country and year. OIL PRODUCING INDUSTRY IN YOUR STATE

Petroleum, domestic imports from the Western Hemisphere by country and year. OIL PRODUCING INDUSTRY IN YOUR STATE

Petroleum, domestic production and new reserves by PAD district. OIL PRODUCING INDUSTRY IN YOUR STATE

Petroleum, domestic production demand, and supply. OIL AND GAS CHEMICAL SERVICE

Petroleum, estimated world reserves by region and country. OIL AND GAS CHEMICAL SERVICE

Petroleum, forecasts of selected statistical data and annual growth rates for the United States. PREDICASTS

Petroleum, forecasts of selected statistical data and annual growth rates for the world, by region and country. WORLD CASTS

Petroleum, freight orginated by U.S. class I railroads by district and commodity, annual. ENERGY STATISTICS

Petroleum, value as a percent of all minerals, by state. OIL PRODUCING INDUSTRY IN YOUR STATE

Petroleum Administration for Defense districts, petroleum statistics by PAD district. TWENTIETH CENTURY PETROLEUM STATISTICS

Petroleum and natural gas, estimated proved reserves. COAL FACTS

Petroleum as percent of all railroad car freight revenue, annual. ENERGY STATISTICS

Petroleum asphalt. See Asphalt (petroleum)

Petroleum bulk plants, total number, sales, and employment. NATIONAL PETROLEUM NEWS FACT BOOK

Petroleum coke, world production by country, annual. WORLD ENERGY SUPPLIES

Petroleum companies, capitalization, income statement, balance sheet, and exploration and production data. OIL AND CHEMICAL GAS SERVICE

Petroleum companies, composite income statement and balance sheet. FINANCIAL ANALYSIS OF A GROUP OF PETROLEUM COMPANIES

Petroleum companies, domestic, estimated crude oil and condensate reserves, by company. OIL AND GAS CHEMICAL SERVICE

Petroleum companies, domestic, ranking of top 100 companies by revenue. OIL AND GAS CHEMICAL SERVICE

Petroleum companies, domestic and foreign net income, annual. BASIC PETROLEUM DATA BOOK

Petroleum companies, domestic refinery runs, net crude oil and natural gas liquids production, and rank in refining, by company. OIL AND GAS CHEMICAL SERVICE

Petroleum companies, foreign, ranking by revenue. OIL AND GAS CHEMICAL SERVICE

Petroleum companies, sources and uses of working capital, annual. BASIC PETROLEUM DATA BOOK

Petroleum companies, sources of internal and external capital, annual. BASIC PETROLEUM DATA BOOK

Petroleum consumption in the United States through the year 2000 by consuming sectors and as a percent of total energy input. BASIC PETROLEUM DATA BOOK

Petroleum (crude), contribution to freight revenue of U.S. railroad companies, annual. ENERGY STATISTICS

Petroleum (crude and products), total transported in the United States, by method of transportation, annual. ENERGY STATISTICS

Petroleum (energy) products, production, trade, and total and per capita apparent consumption, by country, annual. WORLD ENERGY SUPPLIES

Petroleum event, major, international. INTERNATIONAL PETROLEUM

Petroleum fuels, total free-world demand, annual. STANDARD AND POOR'S INDUSTRY SURVEYS

Petroleum geophysical exploration in the United States, annual. BASIC PETROLEUM DATA BOOK

Petroleum industry, distribution of total revenue dollars. FINANCIAL ANALYSIS OF A GROUP OF PETROLEUM COMPANIES

Petroleum industry, earnings reinvested and employed and investment in fixed assets. FINANCIAL ANALYSIS OF A GROUP OF PETROLEUM COMPANIES

Petroleum industry, estimated percentages of company sales, of suppliers of products and services to the industry. STANDARD AND POOR'S INDUSTRY SURVEYS

Petroleum industry, expenditures by function. FINANCIAL ANALYSIS OF A GROUP OF PETROLEUM COMPANIES

Petroleum industry, net income as a percent of net worth compared with other industries. BASIC PETROLEUM DATA BOOK

Petroleum industry, rates of return on invested capital and assets. FINANCIAL ANALYSIS OF A GROUP OF PETROLEUM COMPANIES

Petroleum industry, sources and uses of working capital. STANDARD AND POOR'S INDUSTRY SURVEYS

Petroleum industry, worldwide capital and exploration expenditures, annual. BASIC PETROLEUM DATA BOOK

Petroleum liquids, forecast of selected statistical data and annual growth rate for the United States. PREDICASTS

Petroleum liquids, forecast of selected statistical data and annual growth rate for the world, by region and country. WORLD CASTS

Petroleum movement in the domestic waterborne trade, annual. ENERGY STATISTICS

Petroleum oil transported through domestic pipelines, by company. TRANSPORT STATISTICS

Petroleum products, barge movements from PAD district III to PAD district I via the Mississippi River, daily averages. ANNUAL STATISTICAL REVIEW

Petroleum products, domestic consumption, annual. COAL FACTS

Petroleum products, domestic consumption by state. OIL PRODUCING INDUSTRY IN YOUR STATE

Petroleum products, domestic demand, by type, annual. OIL AND GAS CHEMICAL SERVICE

Petroleum products, domestic exports, imports, and net imports, annual. INTERNATIONAL ECONOMIC REPORT OF THE PRESIDENT

Petroleum products, Federal Reserve Board indexes of industrial production, monthly and annual averages. STANDARD AND POOR'S TRADE AND SECURITIES STATISTICS

Petroleum products, production by selected countries and regions, annual. INTERNATIONAL ECONOMIC REPORT OF THE PRESIDENT

Petroleum products, production imports, exports, consumption by function and user, and consumption for transformation in OECD member countries, annual. STATISTICS OF ENERGY

Petroleum products, transportation by pipelines, and between PAD districts. ENERGY STATISTICS

Petroleum refineries' consumption of natural gas in the United States, annual. GAS FACTS

Petroleum refinery output (all products), by world region and country, annual. WORLD ENERGY SUPPLIES

Petroleum refining industry, hourly wages, weekly earnings, and hours worked, monthly and annual averages. STANDARD AND POOR'S TRADE AND SECURITIES STATISTICS

Petroleum terminals, total number, sales and employment. NATIONAL PETROLEUM NEWS FACT BOOK

Philadelphia (Pa.), distillates and residual fuel oils cargo prices, high, low, and average. PLATT'S OIL PRICE HANDBOOK AND OILMANAC

Philadelphia (Pa.), fuel oil prices. OIL DAILY

Philadelphia (Pa.), gasoline prices, premium, unleaded, and regular. OIL DAILY

Philadelphia (Pa.), motor gasoline refinery and terminal prices, regular and premium. PLATT'S OIL PRICE HANDBOOK AND OILMANAC

Philadelphia (Pa.), propane refinery and terminal prices, high, low, and average. PLATT'S OIL PRICE HANDBOOK AND OILMANAC

Philippines, natural gas reserves, refining capacity, refined products demand, and tank ship fleet. TWENTIETH CENTURY PETROLEUM STATISTICS

Phillips Petroleum Company, crude oil prices (posted). PLATT'S OIL PRICE HANDBOOK AND OILMANAC

Pipeline companies, domestic, assets, liabilities, income account, balance sheet, and operating revenues and expenses of individual companies. TRANS-PORT STATISTICS IN THE UNITED STATES

Pipeline companies, domestic, composite financial and related data, annual. STANDARD AND POOR'S INDUSTRY SURVEYS

Pipeline companies, domestic, composite income account and balance sheet. TRANSPORT STATISTICS IN THE UNITED STATES

Pipeline companies, domestic, employees and their compensation. TRANSPORT STATISTICS IN THE UNITED STATES

Pipeline companies, domestic, miles of main, by company. OIL AND GAS CHEMICAL SERVICE

Pipeline companies, domestic, new construction expenditures, annual. STAN-DARD AND POOR'S INDUSTRY SURVEYS

Pipeline companies, domestic, oil and products originated, transported, and delivered out. TRANSPORT STATISTICS IN THE UNITED STATES

Pipeline companies, domestic, owned miles operated by gathering and trunk lines, by company and by state. TRANSPORT STATISTICS IN THE UNITED STATES

Pipeline companies, domestic, purchases versus production of gas, by company. OIL AND GAS CHEMICAL SERVICE

Pipeline companies, domestic, total number, revenues, expenses, and income, annual. ENERGY STATISTICS

Pipeline companies (interstate natural gas), composite income account and balance sheet. STATISTICS OF INTERSTATE NATURAL GAS PIPELINE COM-PANIES

Pipeline companies (interstate natural gas), composite operating revenue, expenses, and sales, by customer type. STATISTICS OF INTERSTATE NATURAL GAS PIPELINE COMPANIES

Pipeline companies (interstate natural gas), financial and operating statistics on individual companies. STATISTICS OF INTERSTATE NATURAL GAS PIPELINE COMPANIES

Pipeline companies (interstate natural gas), gas utility plant and gas pipeline mileage, by function and pipe. STATISTICS OF INTERSTATE NATURAL GAS PIPELINE COMPANIES

Pipeline companies (interstate natural gas), sources and uses of funds. STATISTICS OF INTERSTATE NATURAL GAS PIPELINE COMPANIES

Pipeline mileage, by type of product, annual. ENERGY STATISTICS

Pipeline mileage in the United States, annual. BASIC PETROLEUM DATA BOOK

Pipelines, capital and exploration expenditures and gross and net investment by the world petroleum industry, by select countries. CAPITAL INVESTMENT IN THE PETROLEUM INDUSTRY

Pipelines, crude oil and petroleum products, average length of movement, annual. ENERGY STATISTICS

Pipelines (interstate), firm requirements and percentage curtailment of gas deliveries. STANDARD AND POOR'S INDUSTRY SURVEYS

Pipeline system, domestic, service interruptions by type. STATISTICS OF INTERSTATE NATURAL GAS PIPELINE COMPANIES

Pipeline terminals, south and southeast motor gasoline refinery and terminal prices, regular and premium. PLATT'S OIL PRICE HANDBOOK AND OIL-MANAC

Plant condensate, imports by the United States, annual. BASIC PETROLEUM DATA BOOK

Poland, crude production, total and as a percent of world production, reserves of crude and natural gas, refining capacity, and tank ship fleet. TWENTIETH CENTURY PETROLEUM STATISTICS

Poland, oil and natural gas production, consumption, and trade. INTERNATIONAL OIL DEVELOPMENTS

Pollution and environmental control revenue bonds, description, face value, term, and rating of individual issues. MOODY'S PUBLIC UTILITY MANUAL

Population served by gas and electric utilities in the United States. FINANCIAL STATISTICS OF PUBLIC UTILITIES

Portland (Oreg.), cement plants consumption of natural gas, annual. GAS FACTS

Portland (Oreg.), motor gasoline refinery and terminal prices, regular and premium. PLATT'S OIL PRICE HANDBOOK AND OILMANAC

Portland, (Oreg.), refinery and terminal prices of distillates and residual fuel oils. PLATT'S OIL PRICE HANDBOOK AND OILMANAC

Portugal, production, consumption, and exports and imports by direction, of crude oil and crude oil products by type, and of natural gas and liquids. QUARTERLY OIL STATISTICS

Portugal, production, consumption, exports, and imports, by function and user, of energy sources by type. STATISTICS OF ENERGY

Portugal, refinery capacity, refined products demand, and tank ship fleet. TWENTIETH CENTURY PETROLEUM STATISTICS

Power bills. See Electric power bills

Power shovels at bituminous coal mines. BITUMINOUS COAL DATA

Preferred stocks issues of gas utility industry by type of company, and cost and yield. GAS FACTS

Preferred stocks of utility companies, new issues, description, face value, rating, price, and yield. MOODY'S PUBLIC UTILITY MANUAL

Premium gasoline, use by passenger cars, annual. NATIONAL PETROLEUM NEWS FACT BOOK

Pressurized heavy-water- and light-water-moderated and cooled reactors, description and net electrical power generation, by country. POWER REACTORS IN MEMBER STATES

Prices. See also under specific items, such as Crude oil prices

Prices of crude and natural gas at well, natural gas at point of consumption, natural gasoline, at plant, price index, posted price by barrel by grades and posted price of major exporting countries. TWENTIETH CENTURY PETROLEUM STATISTICS

Prices of crude oil in OPEC countries. INTERNATIONAL OIL DEVELOPMENTS

Prices of crude oil (posted). See also under specific companies such as Mobil Oil Corporation, crude oil prices (posted)

Prices of retail petroleum products in the United States and other OECD countries in U.S. cents per gallon. INTERNATIONAL OIL DEVELOPMENTS

Primary energy production, world, regional, and by country, annual. WORLD ENERGY SUPPLIES

Primary energy production by selected industrial countries, annual. INTERNATIONAL ECONOMIC REPORT OF THE PRESIDENT

Productivity. See Output

Propane, forecast of selected statistical data and annual growth rates for the United States. PREDICASTS

Propane, forecast of selected statistical data and annual growth rates for the world, by region and country. WORLD CASTS

Propane, futures market delivery date, open, high, low, close, life of contract, and volume of trading. OIL DAILY

Propane, prices in selected cities, high, low, and average. PLATT'S OIL PRICE HANDBOOK AND OILMANAC

Providence (R.I.), distillates and residual fuel oils, cargo prices, high, low, and average. PLATT'S OIL PRICE HANDBOOK AND OILMANAC

Providence (R.I.) motor gasoline refinery and terminal prices, regular and premium. PLATT'S OIL PRICE HANDBOOK AND OILMANAC

Providence (R.I.), refinery and terminal prices of distillates and residual fuel oils. PLATT'S OIL PRICE HANDBOOK AND OILMANAC

Public utilities. See also Utilities

Public utilities, new construction expenditures by type, annual. MOODY'S PUBLIC UTILITY MANUAL; STANDARD AND POOR'S INDUSTRY SURVEYS

Pumped storage plants, reported new capability additions by type of ownership. MOODY'S PUBLIC UTILITY MANUAL

Qatar, crude oil prices, by oil field. PLATT'S OIL PRICE HANDBOOK AND OILMANAC

Qatar, crude oil production in barrels per day. INTERNATIONAL OIL DEVELOPMENTS

Qatar, crude oil production in barrels per day and as a percent of total production. INTERNATIONAL ECONOMIC REPORT OF THE PRESIDENT

Qatar, crude production, total and as a percent of world production, reserves of crude and natural gas and refining capacity. TWENTIETH CENTURY PETROLEUM STATISTICS

Qatar, imports from and exports to the United States and other developed countries, total value. INTERNATIONAL OIL DEVELOPMENTS

Qatar, natural gas liquids production in barrels per day. INTERNATIONAL OIL DEVELOPMENTS

Qatar, oil reserves and production, exports of crude oil and petroleum to the United States, annual. BASIC PETROLEUM DATA BOOK

Quaker State Oil Refining Corporation, crude oil prices (posted). PLATT'S OIL PRICE HANDBOOK AND OILMANAC

Railroad fuel, average price for class I railroads, annual. ENERGY STATISTICS

Railroad-owned mines, production of bituminous coal, annual. BITUMINOUS COAL DATA

Railroad revenue for hauling bituminous coal. COAL FACTS

Railroad revenue from bituminous coal freight traffic. COAL FACTS

Railroads, class I, aggregate capacity of open-top hopper cars owned. COAL TRAFFIC ANNUAL

Railroads, class I, bituminous coal hauled and revenue received by individual railroads. COAL TRAFFIC ANNUAL

Railroads, class I, coal revenue freight received as compared to other commodity freight. COAL TRAFFIC ANNUAL

Railroads, class I, consumption of bituminous coal, annual. BITUMINOUS COAL DATA

Railroads, class I, operating expenses and revenue and capital expenditures. COAL TRAFFIC ANNUAL

Railroads, comparative utilization of open-top hopper cars, by select railroads. COAL TRAFFIC ANNUAL

Railroads, open-top hopper cars owned, new and rebuilt, installed and retired, by leading coal carrier. COAL TRAFFIC ANNUAL

Railroads, rates on U.S. coal freight from mines to port of exit. WORLD COAL TRADE

Railroads, revenue from petroleum freight by product, annual. ENERGY STATISTICS

Railroads and railways, annual sales of electricity. EDISON ELECTRIC INSTITUTE STATISTICAL YEARBOOK

Railroad tank cars (petroleum), total number and mileage, annual. ENERGY STATISTICS

Rate of return on borrowed and invested capital, total assets and fixed assets of petroleum industry. STANDARD AND POOR'S INDUSTRY SURVEYS

Rate of return on electric utilities stocks, composite, annual. MOODY'S PUBLIC UTILITY MANUAL

Rate of return on investment in oil companies, by company. OIL AND GAS CHEMICAL SERVICE

Rate of return on net worth in petroleum and other manufacturing industries. ANNUAL STATISTICAL REVIEW

Rate of return on sales, by major industry group. NATIONAL PETROLEUM NEWS FACT BOOK

Rate of return on stockholders equity, by industry group. NATIONAL PETRO-LEUM NEWS FACT BOOK

Rate of return on stockholders equity for domestic energy and related companies. STANDARD AND POOR'S INDUSTRY SURVEYS

Rate structure of individual utility companies. MOODY'S PUBLIC UTILITY MANUAL

Reactors (nuclear). See also under types, such as Advanced gas-cooled graphite-moderated nuclear reactors

Reactors (nuclear), core characteristics initial fuel loading, average enrichment, refuelling, fuel assemblies, discharge, burn up and fuel power density, by country. POWER REACTORS IN MEMBER STATES

Reactors (nuclear), planned by country, general and core characteristics, and plant system. POWER REACTORS IN MEMBER STATES

Reactors (nuclear), plant systems, reactor vessel, coolant, steam generators, turbines, and means of containment control, by country. POWER REACTORS IN MEMBER STATES

Reactors (nuclear), shut downs by country. POWER REACTORS IN MEMBER STATES

Reactors (nuclear), under construction, core characteristics and plant system, by country. POWER REACTORS IN MEMBER STATES

Reactors (nuclear), under construction, with data on their location, contractors, operators, power output, dates of construction, criticality, connection to grid and commercial operation, country. POWER REACTORS IN MEMBER STATES

Reactors (nuclear), worldwide, type and electrical power, by country. POWER REACTORS IN MEMBER STATES

Reactors (nuclear), worldwide, with data on their location, contractors, operators, power output, dates of construction, criticality, connection to grid and commercial operation, by country. POWER REACTORS IN MEMBER STATES

Red neutral oils, refinery and terminal prices, select regions. PLATT'S PRICE HANDBOOK AND OILMANAC

Refined oil pipelines, total mileage, annual. ENERGY STATISTICS

Refined oils, quantity transported through interstate pipelines, annual. STANDARD AND POOR'S TRADE AND SECURITIES STATISTICS

Refined petroleum products and imports demand. MONTHLY ENERGY REVIEW

Refined petroleum products, demand, by country, daily averages. ANNUAL STATISTICAL REVIEW

Refined petroleum products, demand and consumption in the United States. MONTHLY ENERGY REVIEW

Refined petroleum products, demand by Sino-Soviet area. INTERNATIONAL PETROLEUM

Refined petroleum products, demand by the Eastern Hemisphere. INTERNATIONAL PETROLEUM

Refined petroleum products, demand by the Western Hemisphere. INTERNATIONAL PETROLEUM

Refined petroleum products, exports and reexports by Africa. INTERNATIONAL PETROLEUM

Refined petroleum products, exports and reexports by Asiatic area. INTERNATIONAL PETROLEUM

Refined petroleum products, exports and reexports by North America, Central America, and the Caribbean. INTERNATIONAL PETROLEUM

Refined petroleum products, exports and reexports by Sino-Soviet area. INTERNATIONAL PETROLEUM

Refined petroleum products, exports and reexports by South American countries. INTERNATIONAL PETROLEUM

Refined petroleum products, exports and reexports by the Middle East. INTERNATIONAL PETROLEUM

Refined petroleum products, exports and reexports by the United States. INTERNATIONAL PETROLEUM

Refined petroleum products, exports and reexports by Western Europe. INTER-NATIONAL PETROLEUM

Refined petroleum products, imports by Africa. INTERNATIONAL PETROLEUM

Refined petroleum products, imports by Asiatic area. INTERNATIONAL PETRO-LEUM

Refined petroleum products, imports by North America, Central America, and the Caribbean. INTERNATIONAL PETROLEUM

Refined petroleum products, imports by Sino-Soviet area. INTERNATIONAL PETROLEUM

Refined petroleum products, imports by South American countries. INTERNA-TIONAL PETROLEUM

Refined petroleum products, imports by the Middle East. INTERNATIONAL PETROLEUM

Refined petroleum products, imports by the United States. INTERNATIONAL PETROLEUM

Refined petroleum products, imports by the United States, historical data, annual. OIL PRODUCING INDUSTRY IN YOUR STATE

Refined petroleum products, imports by Western Europe. INTERNATIONAL PETROLEUM

Refined petroleum products, movements for the world, by country. INTERNA-TIONAL PETROLEUM

Refined petroleum products, output by Sino-Soviet area. INTERNATIONAL PETROLEUM

Refined petroleum products, output by the Eastern Hemisphere. INTERNATION-AL PETROLEUM

Refined petroleum products, output by the Western Hemisphere. INTERNATION-AL PETROLEUM

Refined petroleum products, percentage yields from crude oil, annual. BASIC PETROLEUM DATA BOOK

Refined petroleum products, transported in the United States, by method of transportation, annual. ENERGY STATISTICS

Refined products, contribution to freight revenue of U.S. railroads, annual. ENERGY STATISTICS

Refined products, conversion factors from metric tons to U.S. barrels, by country. INTERNATIONAL PETROLEUM

Refined products, demand by country or area. INTERNATIONAL PETROLEUM

Refined products, demand by product and country, exports, imports, production, and stocks. TWENTIETH CENTURY PETROLEUM STATISTICS

Refined products, demand by world area, annual. BASIC PETROLEUM DATA BOOK

Refined products, demand by world region and country. STANDARD AND POOR'S INDUSTRY SURVEYS

Refined products, domestic and world demand, annual. TWENTIETH CENTURY PETROLEUM STATISTICS

Refined products, domestic exports, annual. STANDARD AND POOR'S TRADE AND SECURITIES STATISTICS

Refined products, domestic imports, annual. BASIC PETROLEUM DATA BOOK; STANDARD AND POOR'S TRADE AND SECURITIES STATISTICS

Refined products, domestic imports, stocks, and exports, annual. STANDARD AND POOR'S INDUSTRY SURVEYS

Refined products, domestic imports, total, and as a percent of domestic demand, annual. STANDARD AND POOR'S INDUSTRY SURVEYS

Refined products, domestic stocks, annual. STANDARD AND POOR'S TRADE AND SECURITIES STATISTICS

Refined products, domestic stocks, by location, historical data, annual. ANNUAL STATISTICAL REVIEW

Refined products, exports and reexports by country or area. INTERNATIONAL PETROLEUM

Refined products, exports by country of destination, daily averages. ANNUAL STATISTICAL REVIEW

Refined products, imports by country of origin, daily averages. ANNUAL STATISTICAL REVIEW

Refined products, imports by country or area. INTERNATIONAL PETROLEUM

Refined products, imports by selected countries from Arab and non-Arab sources. INTERNATIONAL ECONOMIC REPORT OF THE PRESIDENT

Refined products, output by country or area. INTERNATIONAL PETROLEUM

Refined products, supply and demand by country or area. INTERNATIONAL PETROLEUM

Refined products, wholesale price index, annual and monthly. PLATT'S OIL PRICE HANDBOOK AND OILMANAC·

Refineries, capital and exploration expenditures and gross and net fixed investments by the world petroleum industry, by select countries. CAPITAL INVESTMENTS IN THE PETROLEUM INDUSTRY

Refineries, domestic, number, capacity, input, and percent yields, annual. ENERGY STATISTICS

Refineries, domestic, number and capacity by state and PAD district. OIL AND GAS CHEMICAL SERVICE

Refineries, domestic, number by company and crude capacity. STANDARD AND POOR'S INDUSTRY SURVEYS

Refineries, domestic, number operating and capacity per day, annual. STANDARD AND POOR'S TRADE AND SECURITIES STATISTICS

Refineries, domestic, and free world, total number and capacity, annual. STANDARD AND POOR'S INDUSTRY SURVEYS

Refineries, number and capacity by country. ANNUAL STATISTICAL REVIEW

Refinery capacity, by country and region, annual. WORLD ENERGY SUPPLIES

Refinery capacity, domestic, total operating and by PAD district. TWENTIETH CENTURY PETROLEUM STATISTICS

Refinery capacity of the world, by country. TWENTIETH CENTURY PETROLEUM STATISTICS

Refinery crude throughputs for the world, by country or area, average annual rate. BP STATISTICAL REVIEW OF THE WORLD OIL INDUSTRY

Refinery products, wholesale prices per gallon by type, annual. NATIONAL PETROLEUM NEWS FACT BOOK

Refinery receipts of crude oil by method of transportation, daily averages. ANNUAL STATISTICAL REVIEW

Refinery runs, domestic, by PAD district, annual. BASIC PETROLEUM DATA BOOK

Refinery utilization in Japan, annual. BP STATISTICAL REVIEW OF THE WORLD OIL INDUSTRY

Refinery utilization in the United States, annual. BP STATISTICAL REVIEW OF THE WORLD OIL INDUSTRY

Refinery utilization in Western Europe, annual. BP STATISTICAL REVIEW OF THE WORLD OIL INDUSTRY

Refinery yields, domestic, by product type, annual. STANDARD AND POOR'S INDUSTRY SURVEYS

Refinery yields, percent shortage for the United States. TWENTIETH CENTURY PETROLEUM STATISTICS

Refinery yields in the United States. INTERNATIONAL PETROLEUM

Refinery yields of crude, percentage. BP STATISTICAL REVIEW OF THE WORLD OIL INDUSTRY

Refinery yields of fuel oil, percentages by weight and volume in Western Europe. BP STATISTICAL REVIEW OF THE WORLD OIL INDUSTRY

Refinery yields of gasoline, percentages by weight and volume in the United States. BP STATISTICAL REVIEW OF THE WORLD OIL INDUSTRY

Refinery yields of gasoline, percentages by weight and volume in Western Europe. BP STATISTICAL REVIEW OF THE WORLD OIL INDUSTRY

Refinery yields of middle distillates, percentages by weight and volume in the United States. BP STATISICAL REVIEW OF THE WORLD OIL INDUSTRY

Refinery yields of middle distillates, percentages by weight and volume in Western Europe. BP STATISTICAL REVIEW OF THE WORLD OIL INDUSTRY

Refinery yields throughout the free world, percentages. INTERNATIONAL PETROLEUM

Refining capacity, crude oil, of selected countries and regions in barrels per day and as a percent of total. INTERNATIONAL ECONOMIC REPORT OF THE PRESIDENT

Refining capacity, domestic, total number of operating and shut down refineries, annual. BASIC PETROLEUM DATA BOOK

Refining capacity, estimates by world area, annual. BASIC PETROLEUM DATA BOOK

Reserves. See also under specific fuel, such as Natural gas reserves

Reserves, estimated, of crude oil and natural gas, by world region and country. INTERNATIONAL OIL DEVELOPMENTS

Residential customers, sales of electric power, annual. STANDARD AND POOR'S INDUSTRY SURVEYS

Residential electricity sales revenue, annual. EDISON ELECTRIC INSTITUTE STATISTICAL YEARBOOK

Residential energy consumption, domestic. MONTHLY ENERGY REVIEW

Residential sales of electricity, annual. EDISON ELECTRIC INSTITUTE STATISTICAL YEARBOOK

Residual fuel oil, domestic demand, annual. NATIONAL PETROLEUM NEWS FACT BOOK; STANDARD AND POOR'S INDUSTRY SURVEYS

Residual fuel oil, domestic demand, exports, imports, production, and percent refinery yield. TWENTIETH CENTURY PETROLEUM STATISTICS

Residual fuel oil, domestic demand, production, imports and stocks. MONTHLY ENERGY REVIEW

Residual fuel oil, domestic demand and supply, by use, daily averages. ANNUAL STATISTICAL REVIEW

Residual fuel oil, domestic production, annual. STANDARD AND POOR'S TRADE AND SECURITIES STATISTICS

Residual fuel oil, domestic production, demand and end-of-the-month stocks. STANDARD AND POOR'S TRADE AND SECURITIES STATISTICS

Residual fuel oil, domestic production, stocks, and demand, annual. STANDARD AND POOR'S TRADE AND SECURITIES STATISTICS

Residual fuel oil, forecast of selected statistical data and annual growth rates for the United States. PREDICASTS

Residual fuel oil, forecast of selected statistical data and annual growth rates for the world, by region and country. WORLD CASTS

Residual fuel oil, free-world supply, percentages. INTERNATIONAL PETROLEUM

Residual fuel oil, prices in selected cities, high, low, and average. PLATT'S OIL PRICE HANDBOOK AND OILMANAC

Residual fuel oil, spot prices in selected cities, high, low, and average. PLATT'S OIL PRICE HANDBOOK AND OILMANAC

Residual fuel oil, wholesale price index, annual and monthly. PLATT'S OIL PRICE HANDBOOK AND OILMANAC

Residual fuel oil consumption in the United States. MONTHLY ENERGY REVIEW

Residual heating oil sales in the United States by state and region. NATION-

AL PETROLEUM NEWS FACT BOOK

Residual petroleum products, estimated consumption by state. OIL PRODUCING INDUSTRY IN YOUR STATE

Retail dealer deliveries of bituminous coal, annual. BITUMINOUS COAL DATA

Retail outlets, branded, of oil companies, total number by state and region, by company. NATIONAL PETROLEUM NEWS FACT BOOK

Rhode Island, installed generating capacity, electricity generation by type of prime mover, and electric energy sold by type of customer. EDISON ELECTRIC INSTITUTE STATISTICAL YEARBOOK

Rhode Island, oil companies' market shares, by company. NATIONAL PETROLEUM NEWS FACT BOOK

Road oil, domestic production, annual. STANDARD AND POOR'S TRADE AND SECURITIES STATISTICS

Romania, crude oil production in barrels per day. INTERNATIONAL OIL DEVELOPMENTS

Romania, crude production, total and as a percent of world production, reserves of crude and natural gas, refinery capacity, and tank ship fleet. TWENTIETH CENTURY PETROLEUM STATISTICS

Romania, oil and natural gas production, consumption, and trade. INTERNATIONAL OIL DEVELOPMENTS

Rotary drilling rigs active, by state. OIL PRODUCING INDUSTRY IN YOUR STATE

Rotary rigs, annual activity in the United States by state, and rotary rig census. BASIC PETROLEUM DATA BOOK

Rotary rigs in operation in the United States. MONTHLY ENERGY REVIEW

Runs to still, daily average for domestic and foreign crude and natural gasoline blended. TWENTIETH CENTURY PETROLEUM STATISTICS

Russia. See USSR.

St. Louis (Mo.), gasoline and fuel oil prices. OIL DAILY

St. Louis (Mo.), motor gasoline refinery and terminal prices, regular and premium. PLATT'S OIL PRICE HANDBOOK AND OILMANAC

St. Louis (Mo.), refinery and terminal prices of distillates and residual fuel oils. PLATT'S OIL PRICE HANDBOOK AND OILMANAC

San Francisco (Calif.), motor gasoline refinery and terminal oil prices, regular and premium. PLATT'S OIL PRICE HANDBOOK AND OILMANAC

San Francisco (Calif.), propane refinery and terminal prices, high, low, and average. PLATT'S OIL PRICE HANDBOOK AND OILMANAC

San Francisco (Calif.), refinery and terminal prices of distillates and residual fuel oils. PLATT'S OIL PRICE HANDBOOK AND OILMANAC

Saskatchewan, crude oil prices by oil field. PLATT'S OIL PRICE HANDBOOK AND OILMANAC

Saturations. See Heating saturations

Saudi Arabia, crude oil prices by oil field. PLATT'S OIL PRICE HANDBOOK AND OILMANAC

Saudi Arabia, crude oil production in barrels per day. INTERNATIONAL OIL DEVELOPMENTS

Saudi Arabia, crude oil production in barrels per day and as a percent of total production. INTERNATIONAL ECONOMIC REPORT OF THE PRESIDENT

Saudi Arabia, crude production, total and as a percent of world production, exports to the United States, posted price of crude per barrel, reserves of crude and natural gas, refining capacity, demand for refined products, and tank ship fleet. TWENTIETH CENTURY PETROLEUM STATISTICS

Saudi Arabia, imports from and exports to the United States and other developed countries, total value. INTERNATIONAL OIL DEVELOPMENTS

Saudi Arabia, natural gas liquids production in barrels per day. INTERNATIONAL OIL DEVELOPMENTS

Saudi Arabia, oil reserves, production, exports to the United States of crude oil and petroleum, annual. BASIC PETROLEUM DATA BOOK

Savannah (Ga.), motor gasoline refinery and terminal prices, regular and premium. PLATT'S OIL PRICE HANDBOOK AND OILMANAC

Savannah (Ga.), refinery and terminal prices of distillates and residual fuel oils. PLATT'S OIL PRICE HANDBOOK AND OILMANAC

Seattle (Wash.), motor gasoline refinery and terminal prices, regular and premium. PLATT'S OIL PRICE HANDBOOK AND OILMANAC

Senegal, tank ship fleet, deadweight tonnage. TWENTIETH CENTURY PETROLEUM STATISTICS

Service stations. See also Gasoline service stations

Service stations, gasoline prices in metropolitan areas. NATIONAL PETRO-
LEUM NEW FACT BOOK

Service stations, new openings, closings, and modernizations. NATIONAL
PETROLEUM NEWS FACT BOOK

Service stations, sales of tires, batteries, and accessories, and sales and
operating ratios by state and city. NATIONAL PETROLEUM NEWS FACT
BOOK

Service wells, number drilled in the United States, annual. STANDARD AND
POOR'S TRADE AND SECURITIES STATISTICS

Shale oil, forecast of selected statistical data and annual growth rates for the
United States. PREDICASTS

Shale oil, forecast of selected statistical data and annual growth rates for the
world by region and country. WORLD CASTS

Sharjah, crude oil production in barrels per day. INTERNATIONAL OIL DE-
VELOPMENTS

Shut ins. See Oil wells, abandoned

Singapore, tank ship fleet, deadweight tonnage. TWENTIETH CENTURY PETRO-
LEUM STATISTICS

Skelly Oil Company, crude oil prices (posted). PLATT'S OIL PRICE HAND-
BOOK AND OILMANAC

Sodium-cooled zirconium-hydride-moderated reactor, description and net
electricity generation, by country. POWER REACTORS IN MEMBER
STATES

Sohio Petroleum Company, crude oil prices (posted). PLATT'S OIL PRICE
HANDBOOK AND OILMANAC

Solid fuels, direction of imports and exports, by country and region, annual.
WORLD ENERGY SUPPLIES

Solid fuels, domestic exports by type, by customs district. WORLD COAL
TRADE

Solid fuels, free-world demand, annual. STANDARD AND POOR'S INDUSTRY
SURVEYS

Solid fuels, production, imports, exports, total and per capita consumption by
country and region, annual. WORLD ENERGY SUPPLIES

Somalia, tank ship fleet, deadweight tonnage. TWENTIETH CENTURY PETRO-LEUM STATISTICS

South Africa, refining capacity, demand for refined products, and tank ship fleet. TWENTIETH CENTURY PETROLEUM STATISTICS

South America, crude production, total and as a percent of world production, demand-production ratio, reserves of crude and natural gas, refining capacity, refined products demand, and tank ship fleet. TWENTIETH CENTURY PETRO-LEUM STATISTICS

South Carolina, installed generating capacity, electricity generation by type of prime mover, and electric energy sold by type of customer. EDISON ELECTRIC INSTITUTE STATISTICAL YEARBOOK

South Carolina, oil companies' market, shares, by company. NATIONAL PETROLEUM NEWS FACT BOOK

South Dakota, crude production, new wells drilled and average daily production per well. TWENTIETH CENTURY PETROLEUM STATISTICS

South Dakota, installed generating capacity, electricity, generation by type of prime mover, and electric energy sold by type of customer. EDISON ELECTRIC INSTITUTE STATISTICAL YEARBOOK

South Dakota, oil and natural gas production, exploration, development, prices, and related data, annual. OIL PRODUCING INDUSTRY IN YOUR STATE

South Dakota, oil companies' market shares, by company. NATIONAL PETRO-LEUM NEWS FACT BOOK

South Dakota, total oil and gas wells, dry holes, development, and exploratory wells drilled, by quarter. QUARTERLY REVIEW OF DRILLING STATISTICS

Southeastern Power Administration, review of operations. MOODY'S PUBLIC UTILITY MANUAL

Southwestern Power Administration, review of operations. MOODY'S PUBLIC UTILITY MANUAL

Spain, crude production, total and as a percent of world production, reserves of crude, refining capacity, demand for refined products, and tank ship fleet. TWENTIETH CENTURY PETROLEUM STATISTICS

Spain, estimated imports of crude oil and refined products from Arab and non-Arab countries by original crude source. INTERNATIONAL OIL DEVELOP-MENTS

Spain, production, consumption, exports, and imports, by function and user, of energy sources by type. STATISTICS OF ENERGY

Spain, production, consumption, exports and imports by direction, of crude oil and crude oil products by type, and of natural gas and liquids. QUARTERLY OIL STATISTICS

Spartanburg (S.C.), motor gasoline refinery and terminal prices, regular and premium. PLATT'S OIL PRICE HANDBOOK AND OILMANAC

Spirit (white), world production by country, annual. WORLD ENERGY SUPPLIES

Spot prices for oil in Western Europe. INTERNATIONAL OIL DEVELOPMENTS

Standard Oil Company of California, crude oil prices (posted). PLATT'S OIL PRICE HANDBOOK AND OILMANAC

State and local serverance taxes paid for oil and gas production, by state. OIL PRODUCING INDUSTRY IN YOUR STATE

States with oil and gas production, area proved productive in acres. OIL PRODUCING INDUSTRY IN YOUR STATE

Steam-electric generating plants, capacity of new conventional plants planned or under construction. STEAM ELECTRIC PLANT FACTORS

Steam-electric plant, capacities by individual utility by state. STEAM ELECTRIC PLANT FACTORS

Steam-electric plant, capacity, net generation, fuel consumption, and unit costs by region and state. COAL FACTS

Steam-electric plant, fuel consumption and unit costs, by region. COAL FACTS

Steam-electric utilities, electricity generation by type of ownership, annual. EDISON ELECTRIC INSTITUTE STATISTICAL YEARBOOK

Steam-electric utilities, installed generating capacity, by ownership, annual. EDISON ELECTRIC INSTITUTE STATISTICAL YEARBOOK

Steam-generating heavy-water reactors, description and net electricity generation, by country. POWER REACTORS IN MEMBER STATES

Steel and other metals used by the gas utility industry. GAS FACTS

Steel and rolling mills, annual consumption of bituminous coal. BITUMINOUS COAL DATA

Steel-company-owned mines, production of bituminous coal, annual. BITUMINOUS COAL DATA

Still gas, domestic production, annual. TWENTIETH CENTURY PETROLEUM STATISTICS

Stocks, domestic, of all oils, crude, natural gas, fuel oil, kerosene, motor fuel, gasoline, and refined products. TWENTIETH CENTURY PETROLEUM STATISTICS

Street and highway lighting, electricity sales, annual. EDISON ELECTRIC INSTITUTE STATISTICAL YEARBOOK

Strip mining, production of bituminous coal, annual. BITUMINOUS COAL DATA

Sun Oil Company, crude oil prices (posted). PLATT'S OIL PRICE HANDBOOK AND OILMANAC

Supply, new annual supply of crude and all oils, domestic. TWENTIETH CENTURY PETROLEUM STATISTICS

Surplus refining capacity over oil consumption in the world. BP STATISTICAL REVIEW OF THE WORLD OIL INDUSTRY

Sweden, production, consumption, imports and exports by direction, of crude oil and crude oil products, by type, and of natural gas and liquids. QUARTERLY OIL STATISTICS

Sweden, production, imports, exports, and consumption, by function and user, of energy sources by type. STATISTICS OF ENERGY

Sweden, refining capacity, demand for refined products, and tank ship fleet. TWENTIETH CENTURY PETROLEUM STATISTICS

Switzerland, production, consumption, imports and exports by direction, of crude oil and crude oil products by type, and of natural gas and liquids. QUARTERLY OIL STATISTICS

Switzerland, production, imports, and consumption, by function and user, of energy sources by type. STATISTICS OF ENERGY

Switzerland, refining capacity and demand for refined products. TWENTIETH CENTURY PETROLEUM STATISTICS

Synthetic crude petroleum, forecast of selected statistical data and annual growth rates for the United States. PREDICASTS

Synthetic crude petroleum, forecast of selected statistical data and annual growth rates for the world, by region and country. WORLD CASTS

Synthetic fuels, forecast of selected statistical data and annual growth rates for the United States. PREDICASTS

Synthetic fuels, forecast of selected statistical data and annual growth rates for the world, by region and country. WORLD CASTS

Synthetic gas and liquids consumption in the United States through the year 2000, by consuming sectors and as a percent of total energy input. BASIC PETROLEUM DATA BOOK

Syria, crude oil production in barrels per day. INTERNATIONAL OIL DEVELOPMENTS

Syria, crude production, total and as a percent of world production, reserves of crude and natural gas, refining capacity, and demand for refined products. TWENTIETH CENTURY PETROLEUM STATISTICS

Tacoma (Wash.), motor gasoline refinery and terminal prices, regular and premium. PLATT'S OIL PRICE HANDBOOK AND OILMANAC

Taiwan, crude production, total and as a percent of world production, reserves of crude, natural reserves, refining capacity, and demand for refined products. TWENTIETH CENTURY PETROLEUM STATISTICS

Tampa (Fla.), motor gasoline, refinery and terminal prices, regular and premium. PLATT'S OIL PRICE HANDBOOK AND OILMANAC

Tampa (Fla.), refinery and terminal prices of distillates and residual fuel oils. PLATT'S OIL PRICE HANDBOOK AND OILMANAC

Tanker, employment in voyages, by country. BP STATISTICAL REVIEW OF THE WORLD OIL INDUSTRY

Tanker and barge movements of crude oil and petroleum products from the U.S. Gulf Coast to the East Coast, daily averages. ANNUAL STATISTICAL REVIEW

Tanker fleet by flag, by country or area. BP STATISTICAL REVIEW OF THE WORLD OIL INDUSTRY

Tanker fleet in the world by age, size, and propulsion. BP STATISTICAL REVIEW OF THE WORLD OIL INDUSTRY

Tanker fleet in the world by flag and ownership. BP STATISTICAL REVIEW OF THE WORLD OIL INDUSTRY

Tanker movements of crude oil and petroleum products from the U.S. Gulf Coast to the West Coast, daily averages. ANNUAL STATISTICAL REVIEW

Tanker rates, weight-average, single voyage, monthly average assessment. PLATT'S OIL PRICE HANDBOOK AND OILMANAC

Tankers, capital and exploration expenditures and gross and net investment, by world petroleum industry, by select countries. CAPITAL INVESTMENTS IN THE PETROLEUM INDUSTRY

Tanker tonnage, laid up or idle in the world. BP STATISTICAL REVIEW OF THE WORLD OIL INDUSTRY

Tank ship fleet, actual U.S. fleet in terms of gross and deadweight tonnage, and average speed, annual. BASIC PETROLEUM DATA BOOK

Tank ship fleet, actual world fleet in terms of gross and deadweight tonnage and average speed. BASIC PETROLEUM DATA BOOK

Tank ship fleet, by age, size, propulsion, and employment. ENERGY STATISTICS

Tank ship fleet, by flag, ownership, and deadweight tonnage, annual. ENERGY STATISTICS

Tank ship fleet, domestic, total number, gross and deadweight tonnage, and average speed, annual. ENERGY STATISTICS

Tank ship fleet, domestic and foreign, total number, gross and deadweight tonnage, and average speed, annual. ENERGY STATISTICS

Tank ship fleet, number, deadweight tonnage, and percent of world fleet, for selected countries. OIL AND GAS CHEMICAL SERVICE

Tank ship fleet (world), age distribution by major flag of registry, annual. BASIC PETROLEUM DATA BOOK

Tank ship fleet (world), by country, average age, speed, deadweight tonnage, gross tonnage, number of ships, and number under construction. TWENTIETH CENTURY PETROLEUM STATISTICS

Tank ship fleet (world), by flag of registry. ANNUAL STATISTICAL REVIEW

Taxes, state and federal, paid by investor-owned utilities. EDISON ELECTRIC INSITUTE STATISTICAL YEARBOOK

Tax rate and collections on gasoline, diesel, and lube oil. NATIONAL PETROLEUM NEWS FACT BOOK

Telephone company stocks, new issues, Moody's weighted averages of yields annual. MOODY'S PUBLIC UTILITY MANUAL

Tennessee, crude production, new wells drilled and daily average production per well. TWENTIETH CENTURY PETROLEUM STATISTICS

Tennessee, installed generating capacity, electricity generation by type of prime mover, and electric energy sold by type of customer. EDISON ELECTRIC INSTITUTE STATISTICAL YEARBOOK

Tennessee, oil and natural gas production, reserves, exploration, development, prices, and related data, annual. OIL PRODUCING INDUSTRY IN YOUR STATE

Tennessee, oil companies' market shares, by company. NATIONAL PETROLEUM NEWS FACT BOOK

Tennessee, total oil and gas wells, dry holes, development, and exploratory wells drilled, by quarter. QUARTERLY REVIEW OF DRILLING STATISTICS

Terminals operated by oil companies, total number. NATIONAL PETROLEUM NEWS FACT BOOK

Texaco Inc., crude oil prices (posted). PLATT'S OIL PRICE HANDBOOK AND OILMANAC

Texas, crude production; new wells drilled and daily average production per well; production of gasoline, natural gas, and liquids; and reserves of crude, natural gas, and liquids. TWENTIETH CENTURY PETROLEUM STATISTICS

Texas, installed generating capacity, electricity generation by type of prime mover, and electric energy sold by type of customer. EDISON ELECTRIC INSTITUTE STATISTICAL YEARBOOK

Texas, oil and natural gas production, reserves, exploration, development, prices, and related data, annual. OIL PRODUCING INDUSTRY IN YOUR STATE

Texas, oil companies' market shares, by company. NATIONAL PETROLEUM NEWS FACT BOOK

Texas, total oil and gas wells, dry holes, development, and exploratory wells drilled, by quarter, by district. QUARTERLY REVIEW OF DRILLING STATISTICS

Thailand, reserves of crude and natural gas, demand for refined products, and tank ship fleet. TWENTIETH CENTURY PETROLEUM STATISTICS

Thermal drying at bituminous coal mines, by state. BITUMINOUS COAL DATA

Thermal-generated electricity, total amount, annual and by state. EDISON ELECTRIC INSTITUTE STATISTICAL YEARBOOK

Tidewater movements of crude oil and products from the Gulf and West Coasts and from the Gulf Coast to the West Coast. ENERGY STATISTICS

Tires, batteries, and accessories, sales at service stations, by state and city. NATIONAL PETROLEUM NEWS FACT BOOK

Tonnage, gross and deadweight, world tank ship fleet. TWENTIETH CENTURY PETROLEUM STATISTICS

Transmission companies, stocks, market price per share, Moody's weighted averages, monthly and annual. MOODY'S PUBLIC UTILITY MANUAL

Transportation. See also Tank ship; Pipeline; Railroads; Motor carriers

Transportation mode of bituminous coal to U.S. destination. COAL TRAFFIC ANNUAL

Transportation sector, consumption of energy in the United States, annual.
BASIC PETROLEUM DATA BOOK

Transportation sector, domestic, energy consumption by type of energy, annual.
ENERGY STATISTICS

Transportation sector, energy consumption in the United States. MONTHLY
ENERGY REVIEW

Trinidad, crude production, total and as a percent of world production, reserves
of crude and natural gas, refining capacity, and demand for refinery products.
TWENTIETH CENTURY PETROLEUM STATISTICS

Trucial states, crude oil production and reserves, total and as a percent of
world production and reserves. TWENTIETH CENTURY PETROLEUM STATISTICS

Truck registrations, domestic, annual. TWENTIETH CENTURY PETROLEUM
STATISTICS

Trucks and buses, registration by state. NATIONAL PETROLEUM NEWS FACT
BOOK

Truck shipments of bituminous coal by origin and destination state. COAL
TRAFFIC ANNUAL

Tunisia, crude production, reserves of crude and natural gas, refining capacity,
demand for refined products, and tank ship fleet. TWENTIETH CENTURY
PETROLEUM STATISTICS

Turkey, crude production, total and as a percent of world production, reserves
of crude and natural gas, refining capacity, and demand for refining products.
TWENTIETH CENTURY PETROLEUM STATISTICS

Turkey, production, consumption, and imports and exports by direction, of crude
oil and crude oil products by type, and of natural gas and liquids. QUARTERLY
OIL STATISTICS

Turkey, production, imports, exports, and consumption, by function and user, of
energy sources by type. STATISTICS OF ENERGY

Underground mining, production of bituminous coal, annual. BITUMINOUS
COAL DATA

Underground storage pools, number of pools and ultimate capacity in the United
States. BITUMINOUS COAL DATA

Unfinished oil, forecast of selected statistical data and annual growth rates for
the United States. PREDICASTS

Unfinished oil, forecast of selected statistical data and annual growth rates for the world by region and country. WORLD CASTS

Unfinished oil, total annual imports by the United States. BASIC PETROLEUM DATA BOOK

Union Oil Company of California, crude oil prices (posted). PLATT'S OIL PRICE HANDBOOK AND OILMANAC

United Arab Emirates, crude oil production in barrels per day. INTERNATIONAL OIL DEVELOPMENTS

United Arab Emirates, imports from and exports to the United States and other developed countries. INTERNATIONAL OIL DEVELOPMENTS

United Arab Emirates, oil reserves and production, exports of crude oil and petroleum to the United States, annual. BASIC PETROLEUM DATA BOOK

United Kingdom, crude oil production, total and as a percent of world production, reserves of crude and natural gas, refining capacity, demand for refined products, and tank ship fleet. TWENTIETH CENTURY PETROLEUM STATISTICS

United Kingdom, crude oil production in barrels per day. INTERNATIONAL OIL DEVELOPMENTS

United Kingdom, crude oil production in barrels per day and as a percent of total production. INTERNATIONAL ECONOMIC REPORT OF THE PRESIDENT

United Kingdom, estimated imports of crude oil and refined products from Arab and Non-Arab countries by original crude source. INTERNATIONAL OIL DEVELOPMENTS

United Kingdom, imports and exports, crude and products. INTERNATIONAL OIL DEVELOPMENTS

United Kingdom, natural gas liquids production in barrels per day. INTERNATIONAL OIL DEVELOPMENTS

United Kingdom, primary energy production, annual. INTERNATIONAL ECONOMIC REPORT OF THE PRESIDENT

United Kingdom, production, consumption, and imports and exports by direction, of crude oil and crude oil products, and of natural gas and liquids. QUARTERLY OIL STATISTICS

United Kingdom, production, imports, exports, and consumption, by function and user, of energy sources by type. STATISTICS OF ENERGY

United States, barge movements of crude oil and petroleum products from PAD districts via the Mississippi river, daily averages. ANNUAL STATISTICAL REVIEW

United States, capital and exploration expenditures, gross and net fixed investments in assets by domestic petroleum industry. CAPITAL INVESTMENT IN THE PETROLEUM INDUSTRY

United States, crude oil production in barrels per day. INTERNATIONAL OIL DEVELOPMENTS

United States, crude oil production in barrels per day and as a percent of total production. INTERNATIONAL ECONOMIC REPORT OF THE PRESIDENT

United States, estimated imports of crude oil and refined products from Arab and Non-Arab countries, by original crude source. INTERNATIONAL OIL DEVELOPMENTS

United States, imports and exports of crude and products. INTERNATIONAL OIL DEVELOPMENTS

United States, natural gas liquids production in barrels per day. INTERNATIONAL OIL DEVELOPMENTS

United States, primary energy production, annual. INTERNATIONAL ECONOMIC REPORT OF THE PRESIDENT

United States, production, consumption, and exports and imports by direction, of crude oil and crude oil products by type, and of natural gas and liquids. QUARTERLY OIL STATISTICS

United States, production, imports, exports, and consumption, by function and user, of energy sources by type. STATISTICS OF ENERGY

United States, refineries, by state and in Puerto Rico, number and capacity. ANNUAL STATISTICAL REVIEW

United States, refineries and percent yields of products, total input, daily averages. ANNUAL STATISTICAL REVIEW

United States, refineries by PAD district, number and capacity. ANNUAL STATISTICAL REVIEW

United States, tanker movements of crude oil and petroleum products from the West Coast to the East Coast, daily averages. ANNUAL STATISTICAL REVIEW

United States Energy Research and Development Administration, review of operations. MOODY'S PUBLIC UTILITY MANUAL

United States Federal Power Commission, review of operations. MOODY'S PUBLIC UTILITY MANUAL

United States Rural Electrification Administration, review of operations.
MOODY'S PUBLIC UTILITY MANUAL

Uranium, domestic production, demand and prices. STANDARD AND POOR'S
INDUSTRY SURVEYS

Uranium, enrichment capacity, delivery commitment of enriched uranium, fuel
fabrication capacity and separative work requirements. OECD URANIUM RE-
SOURCES, PRODUCTION AND DEMAND

Uranium, estimated domestic reserves by state. ENERGY STATISTICS

Uranium, estimated world resources by country, production and capacity, inventories and delivery commitment. OECD URANIUM RESOURCES, PRODUC-
TION AND DEMAND

Uranium, exploration and development drilling, by country, annual. OECD
URANIUM RESOURCES, PRODUCTION AND DEMAND

Uranium, forecast of selected statistical data and annual growth rates for the
United States. PREDICASTS

Uranium, forecast of selected statistical data and annual growth rates for the
world, by region and country. WORLD CASTS

Uranium, world annual and cumulative requirements, annual. OECD URANIUM
RESOURCES, PRODUCTION AND DEMAND

Uranium, world fuel cycle capital requirements. OECD URANIUM RESOURCES,
PRODUCTION AND DEMAND

Uranium, world production by country and region, annual. WORLD ENERGY
SUPPLIES

Uruguay, refining capacity, demand for refinery products and tank ship fleet.
TWENTIETH CENTURY PETROLEUM STATISTICS

USSR, crude oil and natural gas, production, consumption, and trade. INTER-
NATIONAL OIL DEVELOPMENTS

USSR, crude oil production in barrels per day. INTERNATIONAL OIL DEVELOP-
MENTS

USSR, crude production, total and as a percent of world production, reserves of
crude and natural gas, refining capacity, demand for refined products, and tank
ship fleet. TWENTIETH CENTURY PETROLEUM STATISTICS

USSR, natural gas liquids production in barrels per day. INTERNATIONAL OIL
DEVELOPMENTS

USSR, primary energy production, annual. INTERNATIONAL ECONOMIC
REPORT OF THE PRESIDENT

Utah, installed generating capacity, electricity generation by type of prime
mover, and electric energy sold by type of customer. EDISON ELECTRIC
INSTITUTE STATISTICAL YEARBOOK

Utah, natural gas and liquids production, reserves of crude, natural, and
liquids. TWENTIETH CENTURY PETROLEUM STATISTICS

Utah, oil and natural gas production, reserves, exploration, development,
prices, and related data, annual. OIL PRODUCING INDUSTRY IN YOUR
STATE

Utah, oil companies' market shares, by company. NATIONAL PETROLEUM
NEWS FACT BOOK

Utah, total oil and gas wells, dry holes, development, and exploratory wells
drilled, by quarter. QUARTERLY REVIEW OF DRILLING STATISTICS

Utilities. See also Gas and electric utilities; Public utilities

Utilities, Federal Reserve Board indexes of industrial production, monthly and
annual averages. STANDARD AND POOR'S TRADE AND SECURITIES STATIS-
TICS

Utilities, gas and electric, forecast of selected statistical data and annual
growth rates for the United States. PREDICASTS

Utilities, gas and electric, forecast of selected statistical data for the world
by region and country. WORLD CASTS

Utility bonds and notes, chronological list classified by rating. MOODY'S
PUBLIC UTILITY MANUAL

Utility bonds, new issues, Moody's weighted averages of yields, annual.
MOODY'S PUBLIC UTILITY MANUAL

Utility bonds, nonrefundable, description, rating, call price, and interest cost
of individual issues. MOODY'S PUBLIC UTILITY MANUAL

Utility capital, new, Moody's weighted averages of yields on newly issued
domestic bonds and preferred stocks, annual. MOODY'S PUBLIC UTILITY
MANUAL

Utility common stocks, Moody's market price and end-of-the-month weighted
averages, annual and monthly. MOODY'S PUBLIC UTILITY MANUAL

Utility companies, common stock, new issues, description, total offer, price,
and yield of individual issues. MOODY'S PUBLIC UTILITY MANUAL

Utility companies, history, management, business, subsidiaries, properties, rate structure, income statement and balance sheet data, and capitalization of individual companies. MOODY'S PUBLIC UTILITY MANUAL

Utility companies, names of auditors, total number of employees and stockholders, transfer agents, and registrars of individual companies. MOODY'S PUBLIC UTILITY MANUAL

Utility dividend rates, Moody's weighted average, monthly and annual. MOODY'S PUBLIC UTILITY MANUAL

Utility preferred stocks, high and low grade, Moody's yield averages, monthly and annual. MOODY'S PUBLIC UTILITY MANUAL

Utility preferred stocks, new issues, Moody's weighted averages of yields, annual. MOODY'S PUBLIC UTILITY MANUAL

Utility stocks, book value per share at the end of the year, Moody's weighted averages. MOODY'S PUBLIC UTILITY MANUAL

Utility stocks, earnings per share, Moody's weighted averages, monthly and annual. MOODY'S PUBLIC UTILITY MANUAL

Utility stocks, high grade, Moody's preferred stock yield averages, low dividend series, monthly and annual. MOODY'S PUBLIC UTILITY MANUAL

Utility stocks, medium grade, Moody's preferred stock yield averages, monthly and annual. MOODY'S PUBLIC UTILITY MANUAL

Utility stocks, price earnings ratio, Moody's averages, monthly and annual. MOODY'S PUBLIC UTILITY MANUAL

Value at wells. See under specific fuels, such as Natural gas, value at wells

Venezuela, capital and exploration expenditures and gross and net investments in fixed assets, by domestic petroleum industry. CAPITAL INVESTMENTS IN THE PETROLEUM INDUSTRY

Venezuela, crude oil production in barrels per day. INTERNATIONAL OIL DEVELOPMENTS

Venezuela, crude oil production in barrels per day and as a percent of total production. INTERNATIONAL ECONOMIC REPORT OF THE PRESIDENT

Venezuela, crude production, total and as a percent of world production, exports to the United States, posted price per barrel, reserves of crude and natural gas, refining capacity, refinery products demand, and tank ship fleet. TWENTIETH CENTURY PETROLEUM STATISTICS

Venezuela, imports from and exports to the United States and other developed countries. INTERNATIONAL OIL DEVELOPMENTS

Venezuela, natural gas liquids production in barrels per day. INTERNATIONAL OIL DEVELOPMENTS

Venezuela, oil reserves and production, exports of crude oil and petroleum to the United States, annual. BASIC PETROLEUM DATA BOOK

Vermont, installed generating capacity, electricity generation by type of prime mover, and electric energy sold by type of customer. EDISON ELECTRIC INSTITUTE STATISTICAL YEARBOOK

Vermont, oil companies' market shares, by company. NATIONAL PETROLEUM INSTITUTE FACT BOOK

Virginia, crude production, new wells drilled, and daily average production per well. TWENTIETH CENTURY PETROLEUM STATISTICS

Virginia, gasoline and fuel oil prices. OIL DAILY

Virginia, installed generating capacity, electricity generation, by type of prime mover, and electric energy sold by type of customer. EDISON ELECTRIC INSTITUTE STATISTICAL YEARBOOK

Virginia, oil and natural gas production, reserves, exploration, development, prices, and related data, annual. OIL PRODUCING INDUSTRY IN YOUR STATE

Virginia, oil companies' market shares, by company. NATIONAL PETROLEUM NEWS FACT BOOK

Virginia, total oil and gas wells, dry holes, development, and exploratory wells drilled, by quarter. QUARTERLY REVIEW OF DRILLING STATISTICS

Wages and earnings, average hourly in crude oil production, annual, historical data. OIL PRODUCING INDUSTRY IN YOUR STATE

Washington, installed generating capacity, electricity generation, by type of prime mover, and electric energy sold by type of customer. EDISON ELECTRIC INSTITUTE STATISTICAL YEARBOOK

Washington, oil companies' market shares, by company. NATIONAL PETROLEUM NEWS FACT BOOK

Washington, total producing oil wells and daily average production per well. TWENTIETH CENTURY PETROLEUM STATISTICS

Water heaters, automatic, domestic annual manufacturers' shipments. GAS FACTS

Water heaters, domestic annual manufacturers' shipments. MOODY'S PUBLIC UTILITY MANUAL

Water heating fuel, distribution among urban and rural occupied units. GAS FACTS

Water power, as a percent of total U.S. energy resources, annual. STANDARD AND POOR'S TRADE AND SECURITIES STATISTICS

Water power, consumption by the world, by country or area. BP STATISTICAL REVIEW OF THE WORLD OIL INDUSTRY

Water power, developed and undeveloped, by geographic division, annual. ENERGY STATISTICS

Water power, domestic consumption as a percent of total energy consumption. TWENTIETH CENTURY PETROLEUM STATISTICS

Water power, production in the United States, annual. GAS FACTS

Water power, production in the United States, by state. COAL FACTS

Water works companies' stocks, new issues, Moody's weighted average of yields, annual. MOODY'S PUBLIC UTILITY MANUAL

Wax, demand in the United States, annual. TWENTIETH CENTURY PETROLEUM STATISTICS

Wax, domestic demand, production, and percent refinery yield. TWENTIETH CENTURY PETROLEUM STATISTICS

Wax, total production in the United States, annual. STANDARD AND POOR'S TRADE AND SECURITIES STATISTICS

Wax (paraffin), world production by country, annual. WORLD ENERGY SUPPLIES

Well drilling operations, domestic, by type of well, annual. STANDARD AND POOR'S INDUSTRY SURVEYS

Wells. See also Oil wells; Gas wells

Wells, abandoned, average life, completions, well by type, drilling and equipment costs, by state and PAD district. TWENTIETH CENTURY PETROLEUM STATISTICS

Wells, domestic, drilling costs per foot by depth and region, and component costs in drilling and completing wells, annual. STANDARD AND POOR'S INDUSTRY SURVEYS

Wells, drilled, oil, gas, dry, and monthly total footage. MONTHLY ENERGY REVIEW

Wells, drilled in the United States, total number by type of well. ANNUAL STATISTICAL REVIEW

Wells, drilling statistics by classification, annual. BASIC PETROLEUM DATA BOOK

Wells, exploratory total free-world completions, by country. BASIC PETROLEUM DATA BOOK

Wells, footage drilled, and cost of drilling and equipping wells and dry holes, total number annual. ANNUAL STATISTICAL REVIEW

Wells, offshore, drilled in the United States, by state. BASIC PETROLEUM DATA BOOK

Wells, oil, gas, dry holes and offshore, average cost of drilling per well, average depth, and average cost per foot. OIL AND CHEMICAL GAS SERVICE

Wells, oil and gas, forecast of selected statistical data and annual growth rates for the United States. PREDICASTS

Wells, oil and gas, forecast of selected statistical data and annual growth rates for the world, by country and region. WORLD CASTS

Wells, onshore, estimated drilling costs by classification in the United States, annual. BASIC PETROLEUM DATA BOOK

Wells, onshore, total footage drilled and average depth per well, by classification, annual. BASIC PETROLEUM DATA BOOK

Wells, total development wells drilled in the United States, by state and district, by quarter. QUARTERLY REVIEW OF DRILLING STATISTICS

Wells, total exploratory and development wells completed as multiple completions in the United States, by state and district, by quarter. QUARTERLY REVIEW OF DRILLING STATISTICS

Wells, total exploratory wells in the United States by state and district, by quarter. QUARTERLY REVIEW OF DRILLING STATISTICS

Wells, total number, drilling and equipping by depth intervals, estimated costs in the United States. ANNUAL STATISTICAL REVIEW

Wells, total number, footage drilled and estimated drilling costs for oil and gas wells and dry holes, by state. ANNUAL STATISTICAL REVIEW

Wells, total number drilled in the United States by type and average depth per well, annual. STANDARD AND POOR'S TRADE AND SECURITIES STATISTICS

Wells, total number of exploratory drilled in the United States, new field wildcats, oil, gas, and dry, annual. BASIC PETROLEUM DATA BOOK

Wells, total oil and gas wells and dry holes drilled in the United States by state and district, by quarter. QUARTERLY REVIEW OF DRILLING STATISTICS

West Coast gasoline and fuel oil prices. OIL DAILY

Western Europe, capital and exploration expenditures and gross and net investment in domestic petroleum industry. CAPITAL INVESTMENTS IN THE PETROLEUM INDUSTRY

Western Europe, export of solid fuel to other countries by country, annual. WORLD ENERGY SUPPLIES

Western European consumption of Middle East oil, annual. BASIC PETROLEUM DATA BOOK

Western Hemisphere, crude production, total and as a percent of world production, reserves of crude and natural gas, refining capacity, refined products, demand, and tank ship fleet. TWENTIETH CENTURY PETROLEUM STATISTICS

West Texas–New Mexico gasoline and fuel oil prices. OIL DAILY

West Virginia, crude production; new wells drilled; and average daily production per well; production of gasoline, natural gas, and liquids; and reserves of crude, natural gas, and liquids. TWENTIETH CENTURY PETROLEUM STATISTICS

West Virginia, installed generating capacity, electricity generation by type of prime mover, and electric energy sold by type of customer. EDISON ELECTRIC INSTITUTE STATISTICAL YEARBOOK

West Virginia, oil and natural gas production, reserves, exploration, development, prices, and related data, annual. OIL PRODUCING INDUSTRY IN YOUR STATE

West Virginia, oil companies' market shares, by company. NATIONAL PETROLEUM NEWS FACT BOOK

West Virginia, total oil and gas wells, dry holes, development, and exploratory wells drilled, by quarter. QUARTERLY REVIEW OF DRILLING STATISTICS

Wet natural gas, production in the United States by state. COAL FACTS

Wholesale prices of crude and refinery products, by type, annual. NATIONAL PETROLEUM NEWS FACT BOOK

Wilmington (N.C.), motor gasoline refinery and terminal prices, regular and premium. PLATT'S OIL PRICE HANDBOOK AND OILMANAC

Wilmington (N.C.), refinery and terminal prices of distillates and residual fuel oils. PLATT'S OIL PRICE HANDBOOK AND OILMANAC

Wisconsin, gasoline and fuel oil prices. OIL DAILY

Wisconsin, installed generating capacity, electricity generation by type of prime mover, and electric energy sold by type of customer. EDISON ELECTRIC INSTITUTE STATISTICAL YEARBOOK

Wisconsin, oil companies' market shares, by company. NATIONAL PETROLEUM NEWS FACT BOOK

Wood River (III.), propane refinery and terminal prices, high, low, and average. PLATT'S PRICE HANDBOOK AND OILMANAC

Work days, average number, by state and district and by type of mine. BITUMINOUS COAL DATA

Working capital, sources and uses in the domestic petroleum industry. STANDARD AND POOR'S INDUSTRY SURVEYS

Wyoming, crude production; new wells drilled and daily average production per well; production of gasoline, natural gas, and liquids; and reserves of crude, natural gas, and liquids. TWENTIETH CENTURY PETROLEUM STATISTICS

Wyoming, installed generating capacity, electricity generation by type of prime mover, and electric energy sold by type of customer. EDISON ELECTRIC INSTITUTE STATISTICAL YEARBOOK

Wyoming, oil and natural gas production and reserves, exploration, development, prices, and related data, annual. OIL PRODUCING INDUSTRY IN YOUR STATE

Wyoming, oil companies' market shares, by company. NATIONAL PETROLEUM NEWS FACT BOOK

Wyoming, total oil and gas wells, dry holes, development, and exploratory wells drilled, by quarter. QUARTERLY REVIEW OF DRILLING STATISTICS

Yugoslavia, crude production, total and world production, reserves of natural gas and crude, refining capacity, and tank ship fleet. TWENTIETH CENTURY PETROLEUM STATISTICS

Yugoslavia, oil and natural gas production, consumption, and trade. INTERNATIONAL OIL DEVELOPMENTS

Section II
SOURCES ANALYZED

The forty national and international serial sources described in this section contain recurring statistical data on various aspects of energy. The initial section of this book contains their detailed alphabetical keyword/subject analysis. This section gives brief bibliographic information on these serials together with annotations regarding their main focus.

ANNUAL STATISTICAL REVIEW. Washington, D.C., American Petroleum Institute, 1975-- .

> Provides petroleum industry statistics compiled from primary sources which are listed in the bibliography section at the end of the volume. Among other things, the annual contains statistics on the following: liquid hydrocarbon and natural gas reserves, wells and drilling activity, production of crude oil and condensate, imports of crude oil and refined products by destination, total U.S. supply and demand by type of fuel stocks, refinery capacity and crude oil runs, average heating degree days, transportation of energy, prices, labor, and rate of return on investment in petroleum manufacturing industry.

BASIC PETROLEUM DATA BOOK. Washington, D.C.: American Petroleum Institute, 1975-- . Quarterly. Loose-leaf.

> Provides current data on production, demand, prices, sales, investment, imports, refining, consumption by sectors, employment, and research and development expenditures. Primary coverage is given to petroleum industry although considerable information is also provided on natural gas.

BITUMINOUS COAL DATA. Washington, D.C.: National Coal Association, 1948-- . Annual.

> Traces statistically the growth of the American bituminous coal industry. Coverage includes the following areas: production; total value and average per ton; men employed; number of mines; consumption and exports; production by type of mine and mine

size; labor; employment and earnings; accidents and safety; price structure; and existing reserves of anthracite, bituminous coal, coal, crude oil, natural gas, and natural gas liquids.

BP STATISTICAL REVIEW OF THE WORLD OIL INDUSTRY. London: British Petroleum Co., 1956-- . Annual.

This is a review from a major oil company. It contains data on reserves, production, consumption, trade, refining, and tankers. Related statistics include the following: world proven oil reserves; total discovered oil by geographical area; world oil production by country and area; OPEC and non-OPEC production; world oil consumption; interarea movements of oil; imports and exports; world oil supply and demand, refinery capacity and utilization; world tanker fleet by age, size and propulsion; main oil movements by sea; and world primary energy production.

CAPITAL INVESTMENTS IN THE PETROLEUM INDUSTRY. New York: Chase Manhattan Bank, 1955-- . Annual.

Deals with annual capital and exploration expenditures and gross and net investments in fixed assets by American companies. Also includes data on crude oil production, refinery runs, liquid petroleum consumption, and world tanker facilities for petroleum transportation.

COAL FACTS. Washington, D.C.: National Coal Association, 1948-- . Biennial.

Originally called BITUMINOUS COAL FACTS, this is a companion volume to NCA's BITUMINOUS COAL DATA (see above). It is a compendium of current and historical data on all aspects of the coal industry. Specifically, the data include production of mineral fuels; production of electricity from water power and from nuclear power, by state; consumption of energy fuels and electricity; exports of coal and other energy sources from the United States; distribution; fuel and power conversion factors; coal reserves of the United States; and the estimated remaining coal, petroleum, and natural gas reserves of the world.

COAL TRAFFIC ANNUAL. Washington, D.C.: National Coal Association, 1968-- .

Deals with the movement of bituminous coal by mode of transportation, individual carriers, and consumer categories. It also contains data on railroad equipment used in hauling coal, methods by which electric utilities receive their coal, and capital expenditures by class I railroads. Specific information relates to car revenue loadings, revenue freight handled, coal tonnage by railroads by status, weekly, weekly loadings, coal dumpings, and financial aspects of railroads.

EDISON ELECTRIC INSTITUTE STATISTICAL YEARBOOK OF THE ELECTRIC
UTILITY INDUSTRY. New York: Edison Electric Institute, 1931-- .

Contains composite data on the electric utilities, such as total
generating capacity in the United States by ownership and type
of prime mover, power supply in terms of capabilities, peak load,
output, installed capacity, and customer characteristics. Also
given is information on revenue by class, average use and bill
per customer, construction expenditures, and cost trends in electric light and power construction by region.

ENERGY STATISTICS: A SUPPLEMENT TO THE NATIONAL TRANSPORTATION
STATISTICS. Washington, D.C.: U.S. Department of Transportation, 1974-- .
Annual. Available from the Government Printing Office.

Contains selected time series data on the transportation, production, processing, and consumption of energy. The initial sections
cover statistics on energy transportation while the latter part
deals with reserves, production, and refining. Other statistics
provided include proved reserves of crude oil and natural gas,
oil shale and uranium deposits, API refinery capacity survey,
energy consumption, cost of operating automobiles, price of railroad fuel, jet operating expenses, nuclear energy resources, and
regional energy statistics.

FINANCIAL ANALYSIS OF A GROUP OF PETROLEUM COMPANIES. New
York: Chase Manhattan Bank, 1948-- . Annual.

Consists of a composite financial evaluation of U.S. petroleum
companies in terms of income statements, balance sheets, revenue,
operating costs, cash earnings, return on invested capital, sources,
and uses of funds, capital expenditures in the United States and
the rest of the world, dividends and changes in the working
capital, asset distribution, and distribution of total revenue dollars. It also includes selected data on crude oil production, refinery runs, and exploration activities.

FINANCIAL STATISTICS OF PUBLIC UTILITIES. Bloomingdale, Ill.: G.A.
Turner and Associates, 1940-- . Annual.

Contains data on the assets and liabilities, capitalization, long-
term debt, management, area and population served by the
utility, number of customers, and quantity of energy sold by over
two hundred energy utilities in the United States.

GAS FACTS. Washington, D.C.: American Gas Association, 1945-- . Annual.

Reports on historical and current indicators of the gas industry.
Comprehensive statistical information includes the following:
estimated and proved recoverable gas reserves, production and
consumption, export and import, number of gas wells by state,

drillings, new wells, cost of drilling, transmission and distribution, gas utility industry miles of main by type, selected operating statistics of major pipeline systems, marketed production and
interstate shipments of natural gas, financial statistics, price
structure, and customers by type.

INTERNATIONAL ECONOMIC REPORT OF THE PRESIDENT TOGETHER WITH
ANNUAL REPORT OF THE COUNCIL ON INTERNATIONAL ECONOMIC
POLICY. Washington, D.C.: Government Printing Office, 1973-- .

Report covers, among other things, the international energy
scene in terms of world reserves, refining capacity, trade and
consumption, production, imports and exports, and oil prices in
Venezuela, Canada, Persian Gulf, Libya, and Nigeria. Special
attention is given to the impact of oil trade on the United
States.

INTERNATIONAL OIL DEVELOPMENTS: STATISTICAL SURVEY. Washington,
D.C.: U.S. Central Intelligence Agency, 1976-- . Biweekly. Available
from the Documents Expediting Project of the Library of Congress.

Contains comprehensive and current information on the world
crude and natural gas production; OPAEC and OPEC countries'
crude oil production, capacity, and reserves; crude oil imports
by source of selected developed countries; trends in oil trade
of selected developed countries; OECD oil consumption; Western
European oil spot market prices; retail petroleum product prices;
crude oil prices; and selected energy-related data on Communist
countries.

INTERNATIONAL PETROLEUM ANNUAL. Washington, D.C.: U.S. Bureau of
Mines, 1965-- . Available from the Government Printing Office.

This is a review of major international events in the Petroleum
industry. Selected statistical data provided in this volume include the following: crude petroleum production; conversion factors; refinery yields; petroleum movements; output of refined
petroleum products; demand; exports and imports of refined products by country or area; crude oil reserves; producing wells;
refineries and capacity; and retail prices of gasoline, kerosene,
lubricating oils, and bunker fuel oil by country.

MONTHLY ENERGY REVIEW. Washington, D.C.: U.S. Federal Energy Information Administration, 1974-- . Available from the Government Printing Office.

Incorporates the PIMS MONTHLY PETROLEUM REPORT and
MONTHLY ENERGY INDICATORS. Each issue reviews the
significant energy trends during the preceding period. The
bulk of the report is devoted to statistical analysis of coal,

petroleum, natural gas, electricity, nuclear power, and uranium
industries. The analysis takes into account production, consumption, demand, stocks, drilling and exploration activity, marketing, exports and imports, and retail and wholesale price and
cost structure. Selected comparative data on developed countries are also provided.

MOODY'S PUBLIC UTILITY MANUAL. New York: Moody's Investors Services,
1954-- . Annual.

Contains unanalyzed but comprehensive and up-to-date financial
and related information on natural gas and electric utility companies, including detailed history, background, management,
mergers and acquisitions, subsidiaries, construction programs,
principal plants, properties, and rate schedules for each company. The center section is devoted to composite data on
production, distribution, stocks, and reserves of main energy
fuels. The annual is supplemented by the PUBLIC UTILITY
NEWS REPORT, published Tuesdays and Fridays.

NATIONAL PETROLEUM NEWS FACT BOOK. New York: McGraw-Hill,
1909-- . Annual.

Issued as the May issue of the NATIONAL PETROLEUM NEWS,
this is a composite industry-wide report on the following areas:
advertising, brand name sales, capital expenditure, distribution,
stocks, price trends by brands and cities, tax revenue, gasoline
station data, and an analysis of the tires, batteries, and aces-
sories market. Other useful information included in the annual
relates to the Canadian market, imports and exports, labor
statistics, and lists of association and management personnel in
the petroleum industry.

OIL, GAS AND CHEMICAL SERVICE. New York: Kerr & Co. Engineers,
1975-- . Quarterly. Loose-leaf.

This is another good source for operating statistics and the earnings and financial condition of various energy-related companies.
It also contains geological data. The companies included are
oil and natural gas producers, natural gas transporters, and
natural gas distributors. In addition to financial information,
the publication covers the following areas: discoveries, production, reserves, wellhead price, utility sales by class, and refinery runs. Composite data relate to both the American and
selected free world markets.

OIL DAILY. New York: Oil Daily, Inc., 1951-- .

Devoted to analysis of news and trends in the oil industry.

Contains statistical information on the oil futures market, and oil product prices by major states and regions.

PLATT'S OIL PRICE HANDBOOK AND OILMANAC. New York: Platt's Oilgram Price Service, 1923-- . Annual.

Contains comprehensive listing of previous years' oil price data for the petroleum industry, including the following: refinery and terminal prices for motor gasoline, liquefied petroleum gas, and distillates and fuel oils. The gasoline price in fifty-five cities are given, based on the Bureau of Labor Statistics' wholesale price averages. Other information provided in the volume relates to the following: crude oil prices for the United States and Canada, Middle East, Africa, Venezuela, Argentina; AFM barge quotations; European bulk prices and averages; average freight rate assessments; cargo postings; bunkering prices; tanker rates; and cargo price notes.

POWER REACTORS IN THE MEMBER STATES. Vienna, Austria: International Atomic Energy Agency, 1969-- . Annual.

Originally called POWER AND RESEARCH REACTORS IN THE MEMBER STATES, this annual contains statistical information on the various types of nuclear power reactors in member states comprising the IAEA.

PREDICASTS. Cleveland, Ohio: Predicasts, 1973-- . Annual.

This is a most thorough abstracting service covering all forecasts by government, journals, research institutions, and companies on various industries including those which are in energy and energy-related areas. The actual forecasts as well as the source for the various forecasts are given, arranged by an eight-digit standard industry.

QUARTERLY OIL STATISTICS. Paris: Organization for Economic Cooperation and Development, 1961-- .

Superseding the earlier pamphlet-type publication from OECD entitled PROVISIONAL OIL STATISTICS BY QUARTERS, this expanded version provides more accurate and comparable data than the former. The following comprehensive information is included on all advanced nations of the free world: complete balance of energy production, trade, refinery intake, output, and final consumption. There are separate sections on crude oil and feed stocks, consumption for nine product groups, imports by origin, and exports by destination.

QUARTERLY REVIEW OF DRILLING STATISTICS. Washington, D.C.: American Petroleum Institute, 1967-- .

This is a report on drilling activity in the United States in terms of total number of oil and gas wells, footage drilled, and dry holes by state; stratigraphic and core tests; service wells; and multiple completion wells. The appendix lists source agencies for drilling statistics.

RESERVES OF CRUDE OIL, NATURAL GAS LIQUID AND NATURAL GAS IN THE UNITED STATES AND CANADA AND THE UNITED STATES PRODUCTIVE CAPACITY. Washington, D.C.: American Gas Association: American Petroleum Institute; and Canadian Petroleum Institute, 1946-- . Annual.

The data in this report represent the best estimates of reserves of crude oil, natural gas liquids, and natural gas, developed from available information by a committee made up of petroleum engineers and geologists. The data include the following: reserves, production, ultimate recovery, and price--by year of recovery and by geographic area.

STANDARD AND POOR'S INDUSTRY SURVEYS. New York: Standard and Poor Corp., 1964-- . Quarterly.

Up-to-date survey providing financial and market data on American electric utilities, and coal, gasoline, natural gas, offshore drilling, petroleum, and nuclear power companies. Includes information on production, demand, supply, exports and imports, and price structure.

STANDARD AND POOR'S TRADE AND SECURITIES STATISTICS. New York: Standard and Poor Corp., 1926-- . Annual and monthly updates. Loose-leaf.

Contains current and historical data on the production, construction, utilization, exploration, stocks, reserves prices, and imports and exports for all energy and energy-related industries in the United States. The fuels reported on include anthracite, aviation gasoline, jet fuel, bituminous coal, crude petroleum, electricity, fuel oil, gasoline and gas, kerosene, and refined petroleum.

STATISTICS OF ENERGY. Paris: Organization for Economic Cooperation and Development, 1961-- . Annual.

Provides data on the production and uses of energy by sources, imports, exports, and consumption by energy sector. Covers all types of energy including hard coal, coke-oven coke, lignite, and blast furnace gas.

STATISTICS OF INTERSTATE NATURAL GAS PIPELINES. Washington, D.C.: U.S. Federal Power Commission, 1942-- . Annual. Available from the Government Printing Office.

Compiled from statements to the FPC by class A and class B natural gas pipeline companies. It contains data on income accounts and balance sheets for the companies, operating revenues and sales, operating and maintenance expenditures, research and development activities, assets and liabilities, and capitalization. These data are given for the entire industry as well as for individual companies.

STATISTICS OF PRIVATELY OWNED ELECTRIC UTILITIES IN THE UNITED STATES. Washington, D.C.: U.S. Federal Energy Information Administration, 1937-- . Annual. Available from the Government Printing Office.

Contains composite financial and operating ratios, balance sheet and income accounts, operating revenues, sales, maintenance expenditures, generating stations, and electric energy account. Similar current and historical data are provided on individual electric utilities.

STATISTICS OF PUBLICLY OWNED ELECTRIC UTILITIES IN THE UNITED STATES. Washington, D.C.: U.S. Federal Energy Information Administration, 1948-- . Annual. Available from the Government Printing Office.

This is a companion publication to the FPC annual on private utilities (see above). The public utilities volume relates to utilities operated by municipalities and federal power agencies. Selected statistical data include the following: net electric utility generation in the United States, composite balance sheet and income statements, sales and revenue per customer, revenues per kilowatt hour sold, and statements on individual utilities in each state.

STEAM ELECTRIC PLANT FACTORS. Washington, D.C.: National Coal Association, 1950-- . Annual.

Provides geographic data on a number of companies, plants, installed capacity, net generation, fuel units, cost per unit, cost per million BTUs, and geographical comparison of fuel consumption and unit costs. It also gives information on the new steam-electric plants under construction, capacity planned, and new nuclear projects under construction or consideration. Fuel utilization efficiency trends are also analyzed.

TRANSPORT STATISTICS IN THE UNITED STATES. PT. 6. OIL PIPELINES. Washington, D.C.: U.S. Interstate Commerce Commission, 1954-- . Annual. Available from the Government Printing Office.

This report covers total petroleum mileage in the United States in terms of crude-oil-gathering pipelines, crude oil trunk pipelines, and petroleum products pipelines. The data provided include pipeline receipts of crude oil and refined products, number and mileage of privately owned railroad tank cars, taxes paid by oil pipeline companies, and their composite financial ratios.

TWENTIETH CENTURY PETROLEUM STATISTICS. Dallas: DeGolyer & MacNaughten, 1946-- . Annual.

Initially prepared in 1945 by the U.S. Navy, Office of Director of Naval Petroleum and Oil Shale Reserves, this report provides statistical data on petroleum for all the countries of the world, in terms of production, reserves, demand and supply, import and exports, price structure, transportation, tanker movements, rates, and drilling and exploratory activity for new supplies and the cost thereof.

URANIUM RESOURCES, PRODUCTION AND DEMAND. Paris: Organization for Economic Cooperation and Development, 1970-- . Annual.

Published jointly by the OECD Nuclear Agency and the International Atomic Energy Agency, this report gives estimates of proven and provable world resources of uranium, production and productive capacities, requirements, country-by-country review of world nuclear power growth, and annual world uranium requirements for different reactor strategies and power growth.

WORLD CASTS. Cleveland: Predicasts, Inc., 1972-- . Quarterly.

This may be looked upon as a companion volume to PREDICASTS, cited above. While the latter contains energy industry forecasts for the United States, similar data are given in this volume on the different countries of the world, divided into four regions. The format of both the serials is identical, including the listing of industries by an eight-digit industry classification.

WORLD COAL TRADE. Washington, D.C.: National Coal Association, 1970-- . Annual.

Covers international trade between coal-producing and -consuming countries. It contains statistics on the following: world production and reserves, U.S. share of the coal trade and value, output in mines, method of transportation, rail rates on U.S. coal exports, and the role of tidewater ports in the coal export trade.

WORLD ENERGY SUPPLIES. New York: United Nations Statistical Office, Series J, 1958-- . Annual.

Report on the production, trade, and consumption of commercial energy by country. The energy resources dealt with include solid fuels, crude petroleum, petroleum products, gaseous fuels, electrical energy, and nuclear fuels.

Section III

ADDITIONAL SOURCES OF STATISTICAL DATA

This section contains an annotated list of some six hundred additional private as well as government sources of statistical data on energy. It is subdivided into various categories such as coal, electricity, natural gas, nuclear power, petroleum, solar and other power, composite sources, and energy and transportation, for the sake of convenience and easy reference.

COAL

COAL AGE. New York: McGraw-Hill, 1911-- . Monthly.

> Worldwide trends in coal production, marketing, transportation, finance, and pricing. Also contains coal price information by mines and a comparative index of fuel.

Colorado. Geological Survey. COAL MINES OF COLORADO, STATISTICAL DATA. Compiled by David C. Jones and D. Keith Murray. Information Series. Denver, Colo.: 1976.

Energy Modeling Forum. COAL IN TRANSITION: 1980-2000. Stanford, Calif.: Stanford University, 1978.

> Contains occasional statistical information on coal industry.

ENGINEERING AND MINING JOURNAL. New York: McGraw-Hill, 1866-- . Monthly.

> Contains occasional statistical information on coal industry.

Financial Post. SURVEY OF MINES. Toronto: 1938-- . Annual.

> Financial and related statistical data on Canadian mines. Includes coal mines.

Gunwaldsen, D. STUDY OF POTENTIAL COAL UTILIZATION, 1985-2000. Upton, N.Y.: Brookhaven National Laboratory, 1977.

HYDROCARBON PROCESSING. Houston: Gulf Publishing Co., 1922-- . Monthly.

Contains occasional statistical information on coal industry.

Illinois. Department of Mines and Minerals. COAL REPORT. Springfield. Annual.

KEYSTONE NEWS BULLETIN. New York: McGraw-Hill, 1977-- . Monthly.

Contains occasional statistical and financial information on coal industry.

McGraw-Hill Co. KEYSTONE COAL INDUSTRY MANUAL. New York: 1918-- . Annual.

Gives information on coal companies, in addition to the following general statistics related to the industry: exports, employment, bargeline companies, coal reserves by company, coal resources by state, fuels used by electric utilities, regional fuel use by companies, transportation of coal, world coal production, number of mines, method of mining, value, preparation, consumption, stocks and reserves, and top producing companies.

A MARKETER'S GUIDE TO THE COAL INDUSTRY. New York: McGraw-Hill, 1977.

Contains statistical and related data on the coal industry.

Miller-Freeman Publications. WORLD MINE REGISTER. San Francisco: 1881-- . Annual.

Guide to active mining and mineral processing operations. The descriptions include the type of operations, ores, mines, and processes. This is a companion publication to WORLD MINING and WORLD COAL. Another similar publication in the area is the well-known MINING INTERNATIONAL YEARBOOK, from the FINANCIAL TIMES of London, which deals with the principal international companies associated with the mining industry.

MINING CONGRESS JOURNAL. Washington, D.C.: American Mining Congress, 1915-- . Monthly.

Contains occasional statistical information on coal industry.

Missouri. Department of Natural Resources. Division of Research and Technical Information. MINABLE COAL RESERVES OF MISSOURI. By E.E. Robertson. Rolla: 1973.

Noyes, R., ed. COAL RESOURCES: CHARACTERISTICS AND OWNERSHIP IN THE U.S.A. Park Ridge, N.J.: Noyes Data Corp., 1978.

Organization for Economic Cooperation and Development. STEAM COAL--
PROSPECTS TO 2000. Paris: 1978.

>Contains projections of steam coal demand and trade to the year
>2000 in a total energy context. Analyzes constraints to expand-
>ing coal use.

United Nations. Economic Commission for Europe. ANNUAL BULLETIN OF
COAL STATISTICS FOR EUROPE. New York: 1955-- .

>Data on the production, imports, exports, and deliveries of solid
>fuels (hard coal, patent fuel and coke-oven coke, gas coke,
>brown coal), structural production, employment and productivity
>of labor, production of hydraulic energy, deliveries of natural
>gas and petroleum products for inland consumption, and consump-
>tion by type of customer.

_____. QUARTERLY BULLETIN OF COAL STATISTICS. New York: 1958-- .

United Nations. Economic Commission for Europe. THE COAL SITUATION IN
EUROPE AND ITS PROSPECTS. New York: Annual.

>Report on the consumption of total primary energy in the ECE
>region, share of solid fuels and their main competitors, in the
>primary energy market, internal deliveries of hard coal to main
>consuming sectors in Western and Eastern Europe, production of
>solid fuels in the ECE region, rate of increase of output, under-
>ground employment in hard coal mines, European trade in hard
>coal and patent fuel, trade in hard coal in the ECE region, and
>the Western and Eastern hard coal balance.

U.S. Bureau of Mines. COAL TRANSPORTATION PRACTICES AND EQUIP-
MENT REQUIREMENTS TO 1985. Bureau of Mines Information Circular, no.
8706. By Gary M. Larwood and David C. Benson. Washington, D.C.:
Government Printing Office, 1976.

>Analysis of the origin-destination patterns of coal transportation
>and modal shares, primarily for rail and river. Projects the
>amount of equipment that will be needed in 1985.

_____. INTERNATIONAL COAL TRADE. Washington, D.C.: Government
Printing Office, 1931-- . Monthly.

>Report on international trade by country of origin and consumer
>groups like public utilities, iron and steel industry, household
>and small industry. It includes information by country on imports
>of coal, coke, (anthracite and bituminous), briquets, exports,
>consumption by end use and type, and production figures.

_____. RESERVE BASE OF BITUMINOUS COAL AND ANTHRACITE FOR
UNDERGROUND MINING IN THE EASTERN UNITED STATES. Bureau of

Mines Information Circular, no. 8655. Washington, D.C.: Government Printing Office, 1974.

> Deals with the location and extent of coal reserves, minable by underground methods, in eleven coal-producing states east of the Mississippi River. Reserve estimates are made for anthracite and bituminous coal beds at depths of less than 1,000 feet.

_____. THE RESERVE BASE OF U.S. COAL BY SULFUR CONTENT. Bureau of Mines Information Circular, nos. 8680 and 8693. Washington, D.C.: Government Printing Office, 1975.

> Contains a state-by-state analysis of the coal reserves, with special reference to their sulfur content.

_____. SUPPLY AND DEMAND FOR U.S. COKING COALS AND METAL-LURGICAL COKE. Special publication, no. 9-76. By Eugene T. Sheridan and George Markon. Washington, D.C.: 1976.

> Deals with availability of coking coals for coke production in the United States, status of coke industry, and expected requirements for metallurgical coke in the United States through 1985.

U.S. Congress. Senate. Public Works Committee. GREATER COAL UTILIZATION. 94th Cong., 1st sess. Washington, D.C.: Government Printing Office, 1975.

> Focuses on coal resources and production, potential environmental impact of coal-fired generating units, and comparison of rail and pipeline coal transport methods.

U.S. Department of the Interior. Geological Survey Conservation Division. FEDERAL INDIAN LANDS. COAL, POTASH, SODIUM AND OTHER MINERAL PRODUCTION AND ROYALTY INCOME RELATED STATISTICS. Washington, D.C.: Government Printing Office, 1971-- . Annual.

> Contains data on the quantity, value, royalty, production, and revenue of coal and uranium, arranged by products, lands, years, and states. There is a companion publication containing similar data for oil and gas.

U.S. Department of the Interior. Office of Coal Research. PROSPECTIVE REGIONAL MARKETS FOR COAL CONVERSION PLANT PRODUCTS PROJECTED TO 1980 AND 1985. Prepared by Foster Associates. Washington, D.C.: Government Printing Office, 1974.

> Evaluation of the potential market in 1980 and 1985 for liquid and gas coal-conversion products. Contains data on energy supply, demand, consumption, production costs, and prices, by PAD district and four coal mining areas, for both natural and synthetic fuels.

U.S. Energy Information Administration. COAL DATA, A REFERENCE. Washington, D.C.: Government Printing Office, 1978-- . Annual.

> Reports on coal reserves, production, consumption, and trade; and on coal industry employment and productivity.

U.S. Federal Energy Administration. ELECTRIC UTILITY COAL CONSUMP-TION AND GENERATION TRENDS, 1976-85. Report by ICF, Inc. Washington, D.C.: 1976.

> Report on projected electricity generation requirements and capacities, and resulting utility coal consumption, for each of nine Regional Electric Reliability Councils.

_____. SHORT-TERM COAL FORECAST, 1975-80. FINAL REPORT. Report by ICF, Inc. Washington, D.C.: 1975.

> Report estimating bituminous coal and lignite production, consumption, and stock levels at selected rates of electrical generation, 1975-80. Also includes estimated price impacts of projections, and breakdowns by coal supply regions and census divisions.

U.S. Geological Survey. COAL RESOURCES OF THE UNITED STATES, JAN. 1, 1974. Geological Survey Bulletin, no. 1412. By Paul Averitt. Washington, D.C.: Government Printing Office, 1975.

> Estimates identified and undiscovered remaining coal resources of the United States as of January 1974. Resource estimates are presented by state, thickness of overburden, coal rank, and thickness of beds.

ELECTRIC POWER

Acton, Jean Paul. PROJECTED NATIONWIDE ENERGY AND CAPACITY SAVINGS FROM PEAK-LOAD PRICING OF ELECTRICITY IN THE INDUSTRIAL SECTOR. Santa Monica, Calif.: Rand, 1978.

Conference Board. UTILITY APPROPRIATIONS: UTILITY INVESTMENT STA-TISTICS. New York: 1958-- . Quarterly.

> Report on new capital outlay by the nation's utility industry.

Edison Electric Institute. EEI RATE BOOK. New York: 1945-- . Annual.

> Data on residential, commercial, and industrial as well as speciality rates of investor-owned utilities by states and communities.

_____. ELECTRIC POWER SITUATION IN THE UNITED STATES. New York: 1976-- . Semiannual.

A survey of the electric power supply, expansion of electric generating facilities, and the manufacture of heavy electric power equipment in the United States. Covers the entire utility industry including investor-owned systems, those of government agencies, and rural electric cooperatives. Two other updating publications from the institute are the WEEKLY ELECTRIC OUTPUT and the ADVANCE RELEASE OF THE DATA FOR THE STATISTICAL YEARBOOK OF THE ELECTRIC UTILITY INDUSTRY.

_____. HISTORICAL STATISTICS OF THE ELECTRIC UTILITY INDUSTRY. New York: 1962. Annual.

This is a companion publication to the annual STATISTICAL REVIEW. Covers the years from the beginning of this century. It includes investor-owned and other utilities.

_____. POCKETBOOK OF ELECTRIC UTILITY INDUSTRY STATISTICS. New York: 1955-- . Annual.

Statistics on the electric utility industry including capacity, production, capital and finance, sales and revenues, customers, fuels, price, and employment. This is a summary of the EEI STATISTICAL YEARBOOK OF THE ELECTRIC UTILITY INDUSTRY.

ELECTRICAL WORLD. New York: McGraw-Hill, 1874-- . Weekly.

Oriented to electric utility executives, consultants, and regulators, this contains statistical tables covering revenues, rates, deliveries, and sales forecasts for the utility industry.

ELECTRICAL WORLD DIRECTORY OF ELECTRIC UTILITIES. New York: McGraw-Hill, 1904-- . Annual.

Used to be published under different titles such as Powers' CENTRAL STATION DIRECTORY AND BUYERS MANUAL; CENTRAL STATION LIST AND MANUAL OF ELECTRIC LIGHTING; CENTRAL STATION LIST AND BUYERS MANUAL; McGRAW-HILL ELECTRIC DIRECTORY (lighting and power edition); McGRAW-HILL CENTRAL STATION DIRECTORY; and McGRAW-HILL DIRECTORY OF ELECTRIC UTILITIES. Contains composite statistical information on electric utilities in the United States.

THE ELECTRIC LETTER. Mt. Vernon, Ill.: 1973-- . Biweekly.

Information on market research, products and service, research and development, pricing, and advertising and public relations relating to electric utilities.

ELECTRIC VEHICLE NEWS. Westport, Conn.: 1972-- . Irregular.

Florida. Florida Energy Data Center. STATISTICS OF THE FLORIDA ELECTRIC UTILITY INDUSTRY, 1960-1974. Tallahassee: 1975.

HANDY WHITMAN INDEX OF PUBLIC UTILITY CONSTRUCTION COSTS. St. Paul, Minn.: Whitman, Requart and Associates, 1924-- . Semiannual.

    Cost trends for electric and gas utilities covering the United States in six geographical divisions.

Illinois. Commerce Commission. Accounts and Finance. ILLINOIS ELECTRIC UTILITIES: A COMPARATIVE STUDY OF ELECTRIC SALES STATISTICS. Research Bulletin. Springfield: 1931-- . Annual.

Louisiana. Division of Natural Resources and Energy. AN ECONOMIC ATLAS: LOUISIANA'S ELECTRIC UTILITY INDUSTRY. Energy Management Program Report, no. NRE-EM-75-10. By Robert E.C. Weaver and High Thompson. Baton Rouge: 1975.

New York: Department of Public Service. Accounting Systems Section. FINANCIAL STATISTICS OF THE MAJOR PRIVATELY OWNED UTILITIES IN NEW YORK STATE: ELECTRIC, GAS, TELEPHONE, WATER. Albany: 1976-- . Annual.

New York. Department of Public Service. Office of Economic Research. UTILITY STATISTICS HANDBOOK: 1974. Prepared by Andrew Harvey and Julia Bushel. Albany, 1975.

Organization for Economic Cooperation and Development. THE ELECTRICITY SUPPLY INDUSTRY: ANNUAL INQUIRY. Paris: 1967-- . Annual.

    This report gives data on the electric power situation in OECD countries and a short-term forecast.

PUBLIC UTILITIES FORTNIGHTLY. Washington, D.C.: Public Utilities Reports, 1929-- . Biweekly.

    Financial and related data on the utility industry.

United Nations. Economic and Social Commission for Asia and the Pacific. ELECTRIC POWER IN ASIA AND THE PACIFIC. New York: 1955-- . Annual.

    Data on installed generating capacity by ownership and by prime mover, major plants completed, fuel consumption by thermal power stations, electric power supply revenue, and comparative electri-

city bills for domestic consumers, commercial consumers, and industrial consumers.

United Nations. ECONOMIC COMMISSION FOR ASIA AND FAR EAST.
ELECTRIC POWER IN ASIA AND THE FAR EAST. New York: 1955-- .
Annual.

United Nations. Economic Commission for Europe. ANNUAL BULLETIN OF
ELECTRIC ENERGY STATISTICS FOR EUROPE. New York: 1955-- .

Information on maximum net capacity of plants in continuous
operation, production, imports, exports, and consumption of
electric energy, supply of electric energy to consumers, consumption of fuels and corresponding production of electric energy,
and international exchange of electric energy.

U.S. Bureau of Domestic Commerce. ELECTRIC CURRENT ABROAD, 1975.
Washington, D.C.: Government Printing Office, 1975.

Data on manufacturers and exporters, characteristics of electric
current available in principal foreign cities and places throughout
the world. Includes the type of plug used, availability of
adaptors, type and frequency of current, number of phases, nomi-
nal voltage, and frequency stability.

U.S. Department of Energy. INVENTORY OF POWER PLANTS IN THE U.S.
Washington, D.C.: Government Printing Office, 1975-- . Semiannual.

Includes information on generating capacities, primary and alternate fuels, and in-service dates.

U.S. Energy Information Administration. PRINCIPAL ELECTRIC FACILITIES.
Washington, D.C.: Government Printing Office, 1941-- . Annual.

These are a set of 11 oversize maps, with accompanying supple-
mentary report, showing location and capacity of electric utility
transmission lines, generating plants, substations, the location of
submarine and undergound cables, lines under construction, and
interconnections between utilities.

_____. STEAM-ELECTRIC PLANT CONSTRUCTION COST AND ANNUAL
PRODUCTION EXPENSES. Washington, D.C.: Government Printing Office,
1947-- . Annual.

Reports on investments and production costs, fuel use and costs,
capacities, and heat rates of individual nuclear and conventional
steam electric plants in the United States and Puerto Rico.

U.S. Energy Research and Development Administration. ENERGY AND COST

ANALYSIS AND RESIDENTIAL REFRIGERATORS. By Robert A. Hoskins and Eric Hirst. Springfield, Va.: National Technical Information Service, 1977.

> Estimates of the energy and cost impacts of energy-conserving designs for residential electric refrigerators. Includes data on refrigerator heat flows, insulation thickness, food loads, electricity consumption and savings, and costs.

U.S. Federal Energy Information Administration. ALL ELECTRIC HOMES IN THE UNITED STATES. Washington, D.C.: Government Printing Office, 1964-- . Annual.

> Report on electric bills in cities with a population of 50,000 or more. It reviews the trends and levels of national and state bills, trends and levels of community bills, and customer energy use. The report also contains maps of all-electric residential service by size of population.

_____. ANNUAL REPORT. Washington, D.C.: Government Printing Office, 1921-77. Annual.

> Contains, among other things, overviews of the nation's electric power and natural gas industries.

_____. ANNUAL SUMMARY OF COST AND QUALITY OF STEAM-ELECTRIC PLANT FUELS, WITH A SUPPLEMENT ON THE ORIGIN OF THE COAL DELIVERED TO ELECTRIC UTILITIES. Washington, D.C.: 1976-- . Annual.

> Report on the cost and quality of fossil fuel delivered to steam-electric generating plants. Also includes data on delivered coal destination, origin, and selected characteristics, by region.

_____. COST AND QUALITY OF FUELS FOR STEAM-ELECTRIC PLANTS. Washington, D.C.: 1976-- . Monthly.

> Report on the cost and quality of fossil fuels delivered to steam-electric generating plants, having a capacity of at least twenty-five megawatts or with peaking unit plants, representing about 97% of fossil fuel receipts of steam-electric plants.

_____. ELECTRIC POWER STATISTICS. Washington, D.C.: Government Printing Office, 1961-- . Monthly.

> Contains data on water power and fuel production of energy by electric utilities and individual establishments by state and region; electric utilities' consumption of fuels for production of electric energy; electric utility stocks and days supply of coal and oil for the production of electric energy; electric energy for load and peak system load; sales of electric energy to ultimate customers; and sales, revenue and income of class A and B privately owned utilities.

_____. FACTORS AFFECTING THE ELECTRIC POWER SUPPLY, 1980–85. Washington, D.C.: 1976.

Examines the adequacy of electric power supply from 1980 to 1985 in terms of fuel supplies, electric rates, plant construction and maintenance costs, legal restrictions, and industry financing capacity.

_____. GAS TURBINE ELECTRIC PLANT CONSTRUCTION COST AND ANNUAL PRODUCTION EXPENSES. Washington, D.C.: Government Printing Office, 1972.

Gives total number and dollar value of construction expenditure limited to construction of gas turbine electric plants.

_____. HYDROELECTRIC PLANT CONSTRUCTION COST AND ANNUAL PRODUCTION EXPENSES. Washington, D.C., 1957-- .

Report on generating capacities, plant construction and production costs, and hydraulic characteristics of federal and non-federal hydroelectric utility plants in the United States and Puerto Rico.

_____. HYDROELECTRIC POWER RESOURCES OF THE UNITED STATES, DEVELOPED AND UNDEVELOPED. Washington, D.C.: 1953-- . Once every 3/4 years.

This report presents data on the capacity generation and other characteristics of the developed and undeveloped resources in the United States.

_____. NATIONAL ELECTRIC POWER GENERATION AND ENERGY USE TRENDS. Washington, D.C. Quarterly.

Report on trends in the use of energy resources for electric power production, based on electricity utility company reports to the FPC.

_____. NATIONAL ELECTRIC RATE BOOK. Washington, D.C.: Government Printing Office, 1975-- . Updated periodically.

Summaries of rate schedules under which electric service is sold to general ultimate consumers by all privately and publicly owned electric utilities operating in urban areas throughout the United States. The data are compiled from the rate schedules filed with the FPC by electric utilities and companies. Communities having a population of 2,500 or more are covered.

_____. PROPOSED GENERATING CAPACITY ADDITIONS. Washington, D.C.: 1976-- . Annual.

Report on the capacity, number, and types of fossil- and nuclear-fueled generating unit additions planned by electric utilities through 1985.

_____. SALES OF FIRM ELECTRIC POWER FOR RESALE BY PRIVATE ELECTRIC UTILITIES, BY FEDERAL PROJECTS BY MUNICIPALS. Washington, D.C.: Government Printing Office, 1974-- . Annual.

Gives electric wholesale prices and quantities for sales by private utilities and municipalities by power supply regions, and sales by private utilities to total-requirement customers and partial-requirement customers.

_____. A STAFF REPORT OF COST AND QUALITY OF FUELS FOR STEAM ELECTRIC PLANTS. Washington, D.C.: Government Printing Office, 1975.

Describes the volume of coal and gas deliveries by type to companies and plants by state and region, and average prices.

_____. STEAM-ELECTRIC PLANT CONSTRUCTION COST AND ANNUAL PRODUCTION EXPENSES. Washington, D.C.: Government Printing Office, 1974-- . Annual.

Report showing investments and production costs, fuel use and costs, capacities, and heat rates of individual nuclear and conventional steam-electric plants in the United States and Puerto Rico.

_____. SUMMER ELECTRIC LOAD SUPPLY SITUATIONS. Washington, D.C.: Government Printing Office, 1974-- . Annual.

Projection of U.S. load power supply conditions for summer. A similar report is available which projects the winter load. The report also gives data on load situation, scheduled generating capacity, and load supply.

_____. TYPICAL ELECTRIC BILLS. Washington, D.C.: Government Printing Office, 1935-- . Annual.

Survey of typical net monthly bills for residential, commercial, and industrial services. The residential bills cover cities with a population of 2,500 or more. Commercial and industrial bills analyzed refer to bills in cities with a population of 50,000 or more.

_____. WORLD POWER DATA. Washington, D.C.: Government Printing Office. Annual.

Report on the capacity of world electric generating plants and production of electric energy. It gives information on the world electric generating capacity and production, net electric energy production in top ten countries, per capita net electric energy production, installed generating capacity, per capita kilowatt hours production, average annual rates of growth, nuclear power in the world, installed capacity, and annual gross production of nuclear power.

U.S. Federal Energy Information Administration. Bureau of Power. RETAIL ELECTRIC RATE CHANGES. Washington, D.C.: FPC Office of Public Information, 1976-- . Quarterly.

Report on the electric power rate changes.

U.S. Federal Energy Information Administration. Office of Accounting and Finance. PRELIMINARY FIGURES, INCOME, EXPENSES, SALES (CLASS A & B ELECTRIC COMPANIES). Washington, D.C.: FPC Office of Public Information, 1976.

News release containing financial and related data on privately owned electric utilities.

U.S. Geological Survey. LITHIUM RESOURCES AND REQUIREMENTS BY THE YEAR 2000. Geological Survey Professional Paper, no. 1005. Washington, D.C.: Government Printing Office, 1976.

Contains estimates of lithium resources and projected demand. Deals with such potential uses as batteries for energy storage and electric vehicles, fusion reactors, expected oil savings, and research costs. Includes data on U.S. production, shipments, exports, stocks, and prices; and world production capacity, trade, and reserves.

U.S. Rural Electrification Administration. ANNUAL STATISTICAL REPORT. Washington, D.C.: Government Printing Office, 1952-- .

Financial and statistical information about the operations of REA ELECTRIC BORROWERS INDIVIDUALLY AND AS A GROUP. Also included are summary data on cooperative borrowers.

Washington. Utilities and Transportation Commission. Accounting Section. STATISTICS OF ELECTRIC COMPANIES. Olympia: 1935-- . Annual.

Wisconsin. Public Service Commission. Accounts and Finance Division. COMPARISON OF NET MONTHLY BILLS FOR ELECTRIC UTILITY SERVICE IN WISCONSIN. Madison: 1946-- . Annual.

_____. OPERATING REVENUE AND EXPENSE STATISTICS: CLASS A & B PRIVATE ELECTRIC UTILITIES IN WISCONSIN. Madison: 1937-- . Annual.

Wisconsin. Public Service Commission. Rates and Research Division. ANALY-SIS OF OPERATIONS, MUNICIPAL ELECTRIC UTILITIES IN WISCONSIN. Madison: 1942-- . Annual.

NATURAL GAS

American Gas Association. AGA RATE SERVICE. Arlington, Va. Continuous loose-leaf updates.

> Contains gas rate schedules of all types, including current levels of fuel and purchased gas adjustment clauses for gas distribution companies operating in the United States, Canada, and U.S. possessions. Also lists communities supplied with gas, heating value, type of gas served, and rate correspondent.

_____. FORECAST OF CAPITAL REQUIREMENTS OF THE U.S. GAS UTILITY INDUSTRY. Arlington, Va.: 1978.

_____. GAS UTILITY INDUSTRY PROJECTIONS TO 1990. Washington, D.C.: 1966-- . Annual.

> Statistical report on estimated, actual, and projected total gas utility industry sales by region, gas utility sales by type of use, and commercial and industrial gas sales.

_____. GAS UTILITY STATISTICS. Washington, D.C.: 1945-- . Monthly.

> Report on gas sales to residential, commercial, and industrial customers, year-end projections, new residential housing and gas heating, graphic index of total gas sales, monthly gas appliance shipments (i.e., water heating, cooking, drying, and heating equipment).

_____. HISTORICAL STATISTICS OF THE GAS INDUSTRY. 2d ed. Arlington, Va.: 1961. Annual supplements.

> Complete compilation of gas statistics, supplemented by the annual GAS FACTS (see section II).

_____. QUARTERLY REPORT ON GAS INDUSTRY OPERATIONS. Washington, D.C.: 1957-- .

> Provides total industry income statement for the previous twelve months; sales and revenues by industrial, commercial, and residential customers; gas market indicators; regional data; appliance sales; and housing statistics.

_____. U.S. NATIONAL GAS SUPPLY PROJECTIONS, 1976-1990. Arlington, Va.: 1976.

American Petroleum Institute. Division of Statistics. API MONTHLY REPORT COVERING INVENTORIES OF LP-GAS AND LR-GAS. Washington, D.C.: 1976-- .

> Report gives, by geographic regions, the inventories at plants, terminals, refineries, and underground, of the following: ethane, ethylene, propane mix, and isobutaine. There is also a graphic presentation of the data.

BROWN'S DIRECTORY OF NORTH AMERICAN GAS COMPANIES. Duluth, Minn.: Harcourt Brace Jovanovitch, 1887-- . Annual.

> Vital statistics on all private and municipal gas utilities and pipeline companies in the United States and Canada. Contains data on number of customers, employees, kind of gas, average pressure, source of gas, and sales rates.

Economist Intelligence Unit. LNG 1974-1990. Quarterly Economic Review Special Report, no. 17. London: 1974.

> Data on marine operations and market prospects for liquefied natural gas, world trade, carrier capacity, and financing.

Fariday, Edward K. LNG REVIEW--1977. New York: Energy Economics Research, 1978.

Future Requirements Agency. Denver Research Institute, University of Denver, Colorado. FUTURE NATURAL GAS REQUIREMENTS OF THE UNITED STATES. Denver: 1964-- . Annual.

> Continuing series of forecasts of the growth of natural gas requirements in the United States to the year 1990. The growth forecasts are given by class of service by region and state.

Gas Appliance Manufacturers Association. GAS APPLIANCE AND EQUIPMENT SHIPMENTS. New York: 1961-- . Annual.

> Information on energy-using equipment manufacture. Data are given in number of units shipped and exported.

Louisiana. Division of Natural Resources and Energy. 1974 LOUISIANA GAS STATISTICS. Economics and Statistics Report, no. NRE-ES-75-4. By Hilda C. Thornhill and Jennie C. Gooch. Baton Rouge, 1975.

Missouri. Public Service Commission. Office of Economic Research. A REGULATED GAS UTILITY SURVEY: 1865-1974. Jefferson City, 1975.

National LP Gas Association. LP GAS INDUSTRY MARKET FACTS. Oak Brook, Ill.: 1965-- . Annual.

Statistical data on LP gas distribution in the United States.

North Carolina. Interagency Natural Gas Task Force. REPORT ON NATURAL GAS. Raleigh, 1976.

Petroleum Publishing Co. WORLD WIDE REFINING AND GAS PROCESSING DIRECTORY. Tulsa, Okla.: 1947-- . Annual.

> Gives information on the location of U.S. refineries, their capa-cities, and the types of processing. Surveys operating refineries by state and by company within state. Also gives lube; gas capacities; worldwide refining; and world data on oil reserves, wells, oil production, refining capacity, and processing industry construction projects. Lists worldwide refining, gas processing, engineering, and construction companies.

Potential gas Committee. POTENTIAL SUPPLY OF NATURAL GAS IN THE UNITED STATES. Golden, Colo.: 1964-- . Biennial.

> Data on production and estimated reserves of natural gas.

QUARTERLY REPORT OF GAS INDUSTRY OPERATIONS. Arlington, Va.: 1945-- .

> Income statement for industry, plus data on utility consumers, sales, and revenues by type of gas and by region.

United Nations. Economic Commission for Europe. ANNUAL BULLETIN OF GAS STATISTICS FOR EUROPE. New York: 1958-- .

> Data on production and consumption of gas, imports and exports by country, fuels used for production of gas work, number of household consumers of gas, length of mains for transport and distribution of gas, and breakdown of deliveries for fuel consump-tion by type of consumers.

U.S. Bureau of Mines. ANALYSES OF NATURAL GASES. Bureau of Mines Information Circular, no. 8684. By B.J. Moore. Washington, D.C.: Gov-ernment Printing Office, 1961-- . Annual.

> Results of a survey of the composition of 352 natural gas samples from eighteen states and five foreign countires in 1974, as part of an analysis for occurrences of helium in natural gas. Helium surveys of natural gas have been made since 1917 and published annually since 1961.

U.S. Department of the Interior. MAIN LINE NATURAL GAS SALES TO INDUSTRIAL USERS. Washington, D.C.: 1974-- . Annual.

> Deals with natural gas main line (direct) transmission sales to

industrial users, showing quantity and price for individual gas
companies and customers, with summaries by state, two-digit
standard industry code (SIC), and type of sale.

U.S.   Federal Energy Administration.   NATURAL GAS SHORTAGE.   Washington, D.C.:   1975.

Report on the growth of natural gas consumption from 1930
through 1974, reserves from 1947 to 1973, average annual net
reserve additions to interstate and intrastate pipelines, requirements and deficiencies for the largest interstate pipelines, natural
gas deliveries, consumption by sector, and employment in major
gas-consuming industry groups.

U.S.   Federal Energy Information Administration.   FPC QUARTERLY STAFF
REPORTS ON GAS SUPPLY INDICATORS.   Washington, D.C.:   1973-78.

Report of natural gas supply indicators, covering oil and gas
drilling activities, and marketed production and sales by producers
to interstate pipelines by region, from 1971.   Includes historical
data from 1955.

_____.   NATIONAL GAS SUPPLY AND DEMAND 1971-1990.   Washington,
D.C.:   1972.

Contains a forecast of the availability of natural gas and supplemental supplies of gas for a twenty-year period through 1990,
recoverable reserves, production and remaining reserves, historical
and projected demand, and exploratory activity.

_____.   NATIONAL GAS SURVEY.   4 vols.   Washington, D.C.:   1971-75.

Comprehensive study of the present status of the gas industry and
its course of development.   The following are some of the areas
of statistics covered by the publication: gas sales by category,
seasonal residential sales by region, growth of underground storage, ranking of largest companies in the petroleum industry,
concentration ratio in three major gas-producing regions, gas
consumption growth rate, and estimate of ultimate gas reserves.

_____.   NATURAL GAS PIPELINE INTERCONNECTION IN THE UNITED
STATES.   Washington, D.C.:   1977.

Tabulation of current and potential natural gas movement between
areas with shortages and areas with oversupplies.   Also contains
list of plants and fields and their common delivery customers,
pipeline companies showing names and locations of interconnect-
ing companies, current volume of gas receipts and deliveries, and
maximum capability.

_____ . UNDERGROUND STORAGE OF NATURAL GAS BY INTERSTATE PIPELINE COMPANIES. Washington, D.C.: Government Printing Office, 1970-- . Annual.

Report on the number of storage fields, capacity utilization, and certification. Contains data on pipeline companies, estimated cost of storing gas underground, comparison of fixed costs with operating and maintenance expenditures, comparison between the capacity user and the cost of storing underground, gas injections and withdrawals, and comparison of total gas withdrawn from storage.

U.S. Federal Power Commission. Bureau of Natural Gas. THE GAS SUP-PLIES OF INTERSTATE NATURAL GAS PIPELINE COMPANIES. Washington, D.C.: FPC Office of Public Information, 1963-- . Annual.

Statistical summary of the year-end total reserves and delivera-bility, including comparison with past years. It contains maps of natural gas pipelines, summary of domestic natural gas reserves, interstate gas supply composition, production and composite fore-cast of annual deliveries, and deliverability projections to the 1990s.

_____ . U.S. IMPORTS AND EXPORTS OF NATURAL GAS. Washington, D.C.: FPC Office of Public Information, 1976-- . Annual.

Staff release regarding natural gas and liquefied natural gas im-ports and exports by U.S. pipeline companies. The data, derived from company reports to the commission, include the following: summary of U.S. pipeline imports and exports of natural gas, volume and costs of natural gas pipeline imports and exports, volume in BTUs, unit cost in million BTUs, and LNG imports and exports.

U.S. Federal Power Commission. Office of Accounting and Finance. ACTIVI-TIES OF MAJOR PIPELINE COMPANIES. Washington, D.C.: FPC Office of Public Information, 1976-- .

Periodical containing report on financial and statistical data on interstate natural gas pipeline companies.

_____ . ANNUAL REPORT FOR NATURAL GAS PIPELINE COMPANIES (CLASS A, B, C, D) Washington, D.C.: FPC Office of Public Information, 1976-- .

Gives statistical and financial data on natural gas pipeline com-panies.

_____ . AVERAGE PRICES RECEIVED AND PAID BY MAJOR PIPELINES FOR GAS. Washington, D.C.: FPC Office of Public Information, 1976-- . Irregular.

Periodical release on average national monthly sales and purchases, showing the prices, volume, and revenue data.

_____. SALES BY PRODUCERS OF NATURAL GAS TO INTERSTATE PIPE-
LINE COMPANIES. Washington, D.C.: FPC Office of Public Information,
1976.

Report on sales and revenues of all domestic producers and
foreign suppliers by size groups, geographic areas, and by
companies. Include production and related data.

U.S. Office of Technology Assessment. TRANSPORTATION OF LIQUIFIED
NATURAL GAS. Washington, D.C.: Government Printing Office, 1977.

Washington. Utilities and Transportation Commission. STATISTICS OF GAS
COMPANIES. Olympia: 1935-- . Annual.

Wisconsin. Public Service Commission. Accounts and Finance Division.
OPERATING REVENUE AND EXPENSE STATISTICS, CLASS A & B, PRIVATE
GAS UTILITIES IN WISCONSIN. Madison: 1946-- . Annual.

NUCLEAR ENERGY

Atomic Industrial Forum. AIF NUCLEAR FACTSHEET. New York: 1976-- .
Semiannual. Loose-leaf.

Contains statistical data on atomic power, nuclear plants construction, safety, and so forth.

International Atomic Energy Agency. MARKET SURVEY FOR NUCLEAR POWER
IN DEVELOPING COUNTRIES: GENERAL REPORT. New York: UNIPUB,
1973-- . Updated periodically.

Results of a market survey to determine the size and timing of
demand and financial requirements for nuclear power plants in
the 1980s in selected developing countries. Includes summaries
of fourteen country reports. The 1974 edition reevaluates the
potential market and financial requirements for nuclear power in
developing countries in light of increased oil prices.

_____. OPERATING EXPERIENCE WITH NUCLEAR POWER STATIONS IN
MEMBER STATES. Vienna, Austria: 1976-- . Annual.

Report containing performance data, maximum net capacity, the
date the station was commissioned, type of reactor, outrages,
and shutdowns.

Nuclear Assurance Corp. FUEL-TRAC QUARTERLY NUCLEAR INDUSTRY
STATUS. Atlanta: 1973-- .

Contains a listing of nuclear power plants by state, with type of reactor and commercial operation and plants outside the United States, nuclear power capacity by country, recent contract awards, power generation by operating plants, commercial nuclear power generation plant licensing and construction, U.S. fuel requirements, and current inventories of fuel material. This is part of a tripartite publication called FUEL-TRAC QUARTERLY, the other two being the NUCLEAR FUEL STATUS AND FORECAST and NUCLEAR POWER PLANT PERFORMANCE.

NUCLEAR NEWS. Hinsdale, Ill.: American Nuclear Society, 1969-- . Monthly.

Frequent statistical and related information on the nuclear power industry.

U.S. Atomic Energy Commission. NUCLEAR INDUSTRY. Washington, D.C.: Government Printing Office, 1964-- . Annual.

Contains data on shipments of and orders for atomic energy products, employment, investment, exports, nuclear electric power, nuclear components and equipment, reactors, radio isotopes, and radiation applications.

_____. STATISTICAL DATA ON THE URANIUM INDUSTRY. Colorado: Grand Junction Office, 1975-- . Annual.

Now issued by the Energy Research and Development Administration, this is a compilation of historical data on the uranium industry, including resources, ore reserves, production, distribution, drilling, exploration, employment, and commercial sales.

U.S. Atomic Energy Commission. Office of Planning and Analysis. NUCLEAR POWER GROWTH, 1974-2000. Washington, D.C.: Government Printing Office, 1974.

Contains long-term data on nucelar power, breeding, fabrication, fuel cycle, nuclear power plants, plutonium, and reprocessing uranium. There is a supplement to this report, issued in February 1975, entitled TOTAL ENERGY AND ELECTRIC ENERGY AND NUCLEAR POWER PROJECTIONS, U.S.

U.S. Congress. Joint Atomic Energy Committee. FUTURE STRUCTURE OF THE URANIUM ENRICHMENT INDUSTRY. 92d Cong., 1st sess. Washington, D.C.: Government Printing Office, 1973.

Compilation of data on uranium industry in terms of future requirements, supply, costs, and revenue requirements.

U.S. Energy Research and Development Administration. NUCLEAR REACTORS

BUILT, BEING BUILT, OR PLANNED IN THE UNITED STATES. Springfield, Va.: National Technical Information Service, 1976-- . Semiannual.

Lists all nuclear reactors built, under construction, or planned for construction in the United States, by primary function or purpose.

_____. SURVEY OF U.S. URANIUM MARKETING ACTIVITY. Springfield, Va.: National Technical Information Service, 1968-- . Annual.

Periodical report on domestic deliveries, commitments, and requirements; electric utility nuclear fuel arrangements; uranium products, sales agreement; and reactor manufacturers' nuclear fuel purchase agreements. This is a survey of utility companies, reactor manufacturers, and potential uranium producers.

_____. URANIUM EXPLORATION EXPENDITURES IN 1976 AND PLANS FOR 1976-77. Grand Junction Office, Colo.: 1976-- . Annual.

Data on uranium drilling footage, exploration expenditures, and drilling plans. Report of a survey of over one hundred uranium-related companies.

U.S. Energy Research and Development Administration. Division of Production and Materials Management. NUCLEAR INDUSTRY FUEL SUPPLY SURVEY. Oak Ridge, Tenn.: 1975.

Report on the results of a survey of nuclear fuel sales and supply arrangements of civilian utilities, reactor manufacturers, and uranium producers involved in nuclear fuel production and use.

U.S. Energy Research and Development Administration. Division of Reactor Research and Development. OPERATING HISTORY OF U.S. NUCLEAR POWER REACTORS. Oak Ridge, Tenn.: 1974-- .

Contains data on yearly gross generation of electricity for each U.S. nuclear plant. Gives reactor availability and capacity.

U.S. Federal Energy Administration. NET ENERGY FROM NUCLEAR POWER. (Publication no. PB 254 059). By Ralph M. Rotty et al. Springfield, Va.: National Technical Information Service, 1976.

Analyzes net energy costs of electrical power generation from four types of nuclear reactors: boiling water, pressurized water, high temperature gas-cooled, and heavy water. Calculates the energy expenditures for each fuel cycle component.

U.S. Nuclear Regulatory Commission. CONSTRUCTION STATUS OF NUCLEAR POWER PLANTS. Springfield, Va.: National Technical Information Service, 1976-- . Monthly.

Periodical report containing data needed for monitoring the construction for nuclear power plants. Data include construction progress, and current construction and fuel loading schedule revisions for each nuclear power plant currently being built.

_____. OPERATING UNIT STATUS REPORT: LICENSED OPERATING REACTORS. Springfield, Va.: National Technical Information Service, 1976-- . Monthly.

Statistics on U.S. commercial nuclear power reactors relating to power generation, outages, availability, and capacity, as well as comparative performance of fossil and nuclear units. Covers currently licensed and functioning commercial nuclear power plant units.

PETROLEUM

Adelman, Morris A. THE WORLD PETROLEUM MARKET. Baltimore: Published for the Resources for Future by the Johns Hopkins Press, 1972.

Study of the past behavior of oil prices and its future prospects. Data on oil price and cost structure are available. Also given are production and transportation costs which mold price behavior, tanker rates, price movements, price rise chronology, rate of return on equity capital, and debt equity ratio in the oil industry.

American Association of Petroleum Geologists. BULLETIN. Washington, D.C.: 1917-- . Annual June issue.

Contains mostly information like the following: classification of exploratory wells, exploratory test drills in the United States, development wells drilled, wells drilled by state and by type of wells, statistics on exploratory drilling by classification, number and footage drilled in new fields, wildcats for all states, relative success of exploratory drilling, and new oil field discoveries and gas field discoveries grouped by ultimate reserves.

American Petroleum Institute. Division of Statistics. JOINT ASSOCIATION SURVEY. Washington, D.C.: American Petroleum Institute, Petroleum Association of America, and Midcontinent Oil and Gas Association, 1946-- .

Published periodically in two sections. Section 1 includes the estimated cost of drilling oil wells, gas wells, and dry holes, by depth range, for major areas in the United States. Section 2 includes estimated expenditure for oil and gas exploration, development, and production in the United States.

_____. MAGNETIC TAPE FILES FOR WELL AND DRILLING STATISTICS. Washington, D.C.: 1967-- . Irregular.

Data for all wells including development wells, exploratory wells, stratigraphic and core tests, service wells, old wells, and production of crude oil and natural gas.

_____. MONTHLY REPORT ON DRILLING ACTIVITY IN THE UNITED STATES. Washington, D.C.: 1975-- . Monthly.

Shows by state the number of oil wells, gas wells, dry holes, stratigraphic and core tests, and service wells. A table shows exploratory wells by state. Includes a summary of oil wells, gas wells, and dry holes in the development and exploratory categories.

_____. PETROLEUM FACTS AND FIGURES. Washington, D.C.: 1928-71.

Last edition of a discontinued series containing data on production, refining, transportation, marketing and utilization, finance, labor, employment and earnings, fire and safety, well abandonments, finances of petroleum and pipeline companies, capital expenditures, consumption, cost of exploration and development, dealer statistics, geophysical exploration activity, and exports and imports.

_____. REPORTED FIRE LOSSES IN THE PETROLEUM INDUSTRY. Washington, D.C.: 1975-- . Annual.

_____. REVIEW OF FATAL INJURIES IN THE PETROLEUM INDUSTRY. Washington, D.C.: 1975-- . Annual.

_____. SUMMARY OF DISABLING WORK INJURIES IN THE PETROLEUM INDUSTRY. Washington, D.C.: 1975-- . Annual.

_____. SUMMARY OF MOTOR VEHICLE ACCIDENTS IN THE PETROLEUM INDUSTRY. Washington, D.C.: 1975-- . Annual.

_____. WEEKLY STATISTICAL BULLETIN. Washington, D.C.: 1965-- .

Besides a review of the industry trends, the bulletin contains data on the following: input to crude oil distillation units and production of five major products, stocks of motor gasoline, jet fuel (naptha and kerosene type), distillate fuel oil and residual fuel oil, imports of petroleum products, production and stocks of crude oil, and imports of refined products and exports of crude petroleum and refined products from the United States.

Annuaire de L'Europe Petroliere. EUROPEAN PETROLEUM DIRECTORY. Hamburg, W. Germany: 1963-- . Annual.

> Contains data on European petroleum firms and some statistical information.

ARAB OIL AND GAS. Paris: Arab Petroleum Research Center, 1976-- . Fortnightly.

> Records all financial, trade, and industrial developments in the oil- and gas-producing countries, specially Middle East and North Africa.

ARAB OIL AND GAS DIRECTORY. Paris: Arab Petroleum Research Center, 1976-- . Annual.

> Includes comprehensive studies on the current oil and economic situation and future prospects in the Arab countries and Iran.

Association of Oil Pipelines. SHIFTS IN PETROLEUM TRANSPORTATION. Washington, D.C.: 1976-- . Irregular.

> Data on the mode of transportation for crude petroleum and petroleum products, the chief modes being pipelines, water carriers, motor carriers, and railroads.

BUTANE-PROPANE NEWS. Arcadia, Calif.: Butane-Propane News, Inc., 1939-- . Monthly.

> Covers liquefied petroleum gas industry. Includes sales and distribution equipment and appliances.

California. State Lands Commission. ANNUAL AND CUMULATIVE PRODUCTION OF OFFSHORE WELLS IN TIDELAND OIL FIELDS IN CALIFORNIA. Sacramento: 1976-- . Annual.

Canadian Petroleum Association. STATISTICAL YEARBOOK. Calgary, Alberta: 1955-- . Annual.

> Data on Canadian petroleum and natural gas industry, including pipelines, refineries, and processing plants.

Chamber of Commerce of the United States. EMPLOYEE BENEFITS. Washington, D.C.: 1947-- . Biennial.

> Petroleum industry total fringe benefit payments by type of payment.

Chase Manhattan Bank. THE PETROLEUM SITUATION. New York: 1962-- . Monthly.

Review of the demand for petroleum products, domestic production and supply, and wholesale and retail gasoline prices. There is a quarterly and annual summary statistical table covering the demand, supply, and stocks of gasoline, kerosene, distillate, and residual oils.

Chemical Specialties Manufacturers Association. SURVEY OF ANTI-FREEZE SALES. Washington, D.C.: 1976-- . Annual.

Colorado. Geological Survey. OIL AND GAS FIELDS OF COLORADO, STATISTICAL DATA. Compiled by David C. Jones and D. Keith Murray. Information Series. Denver: 1976.

DIRECTORY OF LAND DRILLING AND OILWELL SERVICING CONTRACTORS. Tulsa, Okla.: Petroleum Publishing Co., 1977.

Data on bonafide contractors who drill and perform downhole services on land in the United States, Canada, and overseas. Includes key men, addresses, phone numbers, rated drilling, and depth of rigs.

EASTERN HEMISPHERE PETROLEUM DIRECTORY. Tulsa, Okla.: Petroleum Publishing Co., 1977.

Data on more than 14,000 employees of 5,000 companies in 4,500 locations. Contains names of petroleum firms and those closely related which are active in Europe, Africa, Middle East, and the Asia-Pacific, and brief descriptions of their activity. Companies are listed by country.

Economist Intelligence Unit. OIL PRODUCTION, REVENUES AND ECONOMIC DEVELOPMENT. Quarterly Economic Review Special Report, no. 18. London: 1975.

Data on Iran, Iraq, Kuwait, United Arab Emirates, Oman, Qatar, and Behrain. Forecasts the accumulation of petro-dollars to 1980 and the amount that will be spent outside by each oil-producing state.

Economist Intelligence Unit. OIL REVIEWS. London: 1952-- . Quarterly.

The EIU publishes quarterly economic reviews for all countries, including energy-producing countries. The reviews contain data on production, refining capacity, exports, and revenues. In addition to this, there are five research-based reviews covering the world's main oil-producing areas. These record developments in oil and gas industries and their significance for specific countries concerned and the international oil industry as a whole. They contain numerous statistical data on the industry. The five oil reviews are (1) OIL IN THE FAR EAST AND AUSTRALIA, (2) OIL IN LATIN AMERICA AND THE CARIBBEAN, (3) OIL IN THE MIDDLE EAST, (4) OIL IN NORTH AMERICA, and (4) OIL IN WESTERN EUROPE.

Economist Intelligence Unit. SOVIET OIL TO 1980. Quarterly Economic Review Special Report, no. 14. London: 1973.

Compiled from Russian sources, this report gives data on the Soviet bloc's reserves, production, distribution, domestic consumption, international trade in crude oil and products, and refining capacity.

ELECTRIC HEAT AND AIR CONDITIONING. Chicago: Barks Publications, 1955-- . Monthly.

Data on sales of automatic domestic central heating equipment in the United States. Available from the publisher, 360 North Mitchell, Chicago, Illinois.

ENERGY MANAGEMENT REPORT. Dallas: Energy Communications, 1948-- . Monthly.

Financial data on oil and natural gas industry. Available from the publisher, 800 Davis Building, Dallas, Texas 75202.

THE EUROPEAN OFFSHORE OIL AND GAS YEARBOOK. New York: UNI-PUB, 1977-- . Annual.

Contains data on energy patterns, projections, policy, financing, and investment outlook.

EUROPEAN PETROLEUM YEAR BOOK. New York: International Publications Service, distributors, 1976-- .

Handbook for the European oil and natural gas industry, containing oil and natural gas statistics for individual Western and Eastern European countries. It also provides information on international oil companies operating in Europe; country-by-country details of oil and gas trading and distributing and pipe-line operating companies; storage and transport companies; and professional associations and institutes. It includes a suppliers directory and buyer's guide to products and services.

EXXON, U.S.A. Houston: 1945-- . Quarterly.

Frequent statistical information on petroleum industry.

Financial post. SURVEY OF OIL. Toronto: 1929-- . Annual.

Financial data on petroleum companies listed on the Canadian stock exchanges.

Florida. Energy Data Center. MONTHLY FLORIDA CRUDE OIL AND NATURAL GAS PRODUCTION, SEPT. 1943–MARCH 1975. Tallahassee: 1975.

FUEL OIL AND OIL HEAT. Cedar Grove, N.J.: Industry Publications, 1922-- . Monthly.

Data on sales of automatic domestic central heating equipment, sales of commercial oil heating installations and domestic oil burners. Available from the publisher, 200 Commerce Road, Cedar Grove, New Jersey 07009.

Houthakker, Hendrik S. THE WORLD PRICE OF OIL: A MEDIUM TERM ANALYSIS. Washington, D.C.: American Enterprise Institute for Public Policy Research, 1976.

HUTTLINGER'S PIPELINE REPORT. Washington, D.C.: LaMotte News Bureau, 1960-- . Weekly.

Data on new filings and pipeline applications submitted to the Federal Power Commission by oil and pipeline operators, with information on construction, cost, new construction, and contracts.

Illinois. State Geological Survey. PETROLEUM INDUSTRY IN ILLINOIS. Urbana: 1968-- . Annual.

Indiana. Geological Survey. OIL DEVELOPMENT AND PRODUCTION IN INDIANA. Mineral Economics Series. Bloomington: 1955-- . Annual.

International Energy Agency. OIL MARKET INFORMATION SYSTEM. Paris: 1978.

This is the OECD data base which collects information on all aspects of international petroleum market.

International Oil Scouts Association. INTERNATIONAL OIL AND GAS DEVELOPMENT. Vol. 1: Exploration; vol. 2: Production. Austin, Tex.: 1930-- . Annual.

Reviews geological and geophysical prospecting and leasing activities, wildcat explorations, and proven field development. Also included are exploratory completions, new fields, core drill prospecting, oil and gas production by fields, gas pipeline construction, refining operation, gasoline, and carbon black recycling and repressuring plants.

INTERNATIONAL PETROLEUM ENCYCLOPEDIA. Tulsa, Okla.: Petroleum
Publishing Co., 1975-- . Annual.

> Worldwide survey of crude oil production, imports, reserves,
> natural gas production and imports, refinery capacity and
> throughput, index of oil and gas fields, directory of interna-
> tional oil companies, guide to oil equipment manufacturers,
> suppliers and service companies operating internationally, and
> world's major pipeline contractors.

Interstate Oil Compact Commission. NATURAL STRIPPER WELL SURVEY.
Oklahoma City, Okla.: Issued in cooperation with the National Stripper
Well Association, 1975.

LATEST RESIDUAL FUEL OIL DEVELOPMENTS. New York: National Economic
Research Associates, 1976-- . Monthly.

> Data on supply and demand condition in the residual fuel oil
> market.

LATIN AMERICAN PETROLEUM DIRECTORY. Tulsa, Okla.: Petroleum Publishing Co., 1977-- . Annual.

> Data on more than 4,700 personnel, 1,000 companies, 2,800
> locations. Includes petroleum firms and some closely related
> firms active in Latin America. Companies and activities are
> listed by country. Contains descriptive profiles of petroleum
> activity for countries and sections of Latin America and detailed
> statistical tables showing refining and production for each coun-
> try.

Louisiana. Department of Conservation. LOUISIANA OFFSHORE PETROLEUM
PRODUCTION, BY ZONES. Baton Rouge: 1976-- . Annual.

Louisiana. Division of Natural Resources and Energy. 1974 LOUISIANA
CRUDE OIL STATISTICS. Economic Statistics Report, no. NRE 74-5. By
Paul H. McGinnis and Hild C. Thornhill. Baton Rouge: 1975.

Louisiana. Geological Oil and Gas Division. SUMMARY OF FIELD STATIS-
TICS AND DRILLING OPERATIONS. Baton Rouge: 1976.

Lundberg Survey Inc. THE LUNDBERG LETTER. North Hollywood, Calif.:
1976-- . Weekly.

> Contains charts and tables on the U.S. petroleum industry. In-
> cludes data on gasoline sales, service stations, grade mix, and
> wholesale and retail prices.

Michigan. Geology Division. MICHIGAN'S OIL AND GAS FIELDS: STA-

TISTICAL SUMMARY: DRILLING STATISTICS, PRODUCTION, EXPORTS AND
IMPORTS. Lansing: 1945-- . Annual.

Nebraska. Oil and Gas Conservation Commission. NEBRASKA OIL ACTIVITY
SUMMARY. Lincoln: 1976-- . Annual.

OCEAN CONSTRUCTION AND ENGINEERING REPORT. Houston: Sheffer
Co., 1974-- . Biweekly.

> Data on pipelines, offshore tanker terminals, offshore nuclear
> power plants, seafloor mining, undersea soil mechanics, subsea
> oil storage, orders, contracts, shipyard orders, expansions, and
> closings.

OCEAN OIL WEEKLY REPORT. Houston: Petroleum Publishing Co., 1966-- .
Weekly.

> Data on ocean projects, undersea developments, and firms in-
> volved with offshore oil exploration including stock information,
> earnings reports, mobile rig construction, and offshore activities.

OFFSHORE CONTRACTORS AND EQUIPMENT DIRECTORY. Tulsa, Okla.:
Petroleum Publishing Co., 1977.

> Data on more than 6,500 personnel, 1,200 companies, 2,800
> locations. Provides worldwide coverage of the offshore industry
> in five sections: (1) drilling contractors and owners, (2) con-
> struction contractors, (3) geophysical companies, (4) diving con-
> tractors, and (5) transportation companies.

OFFSHORE RIG LOCATION REPORT. Houston: Offshore Rig Data Services,
1974-- . Monthly.

> Data on worldwide offshore drilling fleets, mobile rigs, and plat-
> form rigs, including worksite, water depth, contractors and shore
> operating base.

OIL AND GAS JOURNAL. Tulsa, Okla.: Petroleum Publishing Co., 1902-- .
Weekly.

> Regularly features industry statistics in addition to annual directory
> and forecast issues. Some of the data available include the fol-
> lowing: API report on crude oil stocks by origin, API report on
> production, Hughes rig count, API refinery report, average pro-
> duction and stocks, suggested prices for major gasoline brands,
> by major city refined product spot prices, imports of crude and
> products, worldwide crude oil and gas production, drilling pro-
> cessing, production, and exploration.

OIL AND PETROLEUM INTERNATIONAL YEARBOOK. London: 1910-- .
Annual.

Continues OIL AND PETROLEUM YEARBOOK and provides statistical and related data on the world petroleum industry.

OIL DAILY. Chicago: 1951-- .

Statistics on oil companies, supplies, imports, production, tanker rates and prices. Available from Oil Daily, 59 East Van Buren, Chicago, Illinois.

OIL DAILY'S OIL INDUSTRY U.S.A. New York: 1976.

Data on individual oil companies and how they have been managed since World War II. Includes histories of three hundred companies with comments from their top executives.

OIL ENERGY STATISTICS BULLETIN. Babson Park, Mass.: Oil Statistics Co., 1923-- . Biweekly.

Statistical analysis of worldwide energy developments and financial and stock market data.

OIL INDUSTRY COMPARATIVE APPRAISALS. Greenwich, Conn.: John S. Herald, Inc., 1948-- . Monthly.

OIL STATISTICS BULLETIN. Babson Park, Mass.: Oil Statistics Co., 1923-- . Biweekly.

Mostly financial and stock market data.

Organization for Economic Cooperation and Development. HISTORICAL OIL STATISTICS. 2 vols. Paris: 1977.

Covering the period between 1960 and 1975, this report provides data on supply and demand for oil in OECD countries.

_____. OIL STATISTICS--STATISTIQUES PETROLIERES. Paris: 1965-- . Annual.

This report contains the following energy data for all the OECD countries: production, supply, disposals, imports by sources, imports from member countries, exports by destination, consumption by end use sectors, processing of crude oil, semiprocessed feed stocks, natural gas, refinery output, and consumption of main petroleum products.

_____. OIL STATISTICS: SUPPLY AND DISPOSAL. Paris: 1965-- . Annual.

Provides data on demand and supply of oil for OECD countries.

Includes crude oil, feed stocks, NGL, and natural gas and petroleum products by category.

_____. PROVISIONAL OIL STATISTICS BY QUARTERS. Paris: 1965-- . Quarterly.

This pamphlet-type publication is superseded now by the comprehensive QUARTERLY REVIEW OF OIL STATISTICS (see below).

_____. QUARTERLY REVIEW OF OIL STATISTICS. Paris: 1977-- .

Supersedes PROVISIONAL OIL STATISTICS BY QUARTERS (see above). Contains data on crude oil and products, including production, refining output, trade, tankers, stocks, and imports and exports by origin and destination.

Organisation of Petroleum Exporting Countries. ANNUAL REVIEW AND RECORD. Vienna: 1968-- .

Surveys major developments in the world oil industry during the review year, with special reference to the OPEC countries. It records important events and activities of the OPEC and its secretariat in Vienna and also statistical data. More detailed information is available on OPEC countries from the annual OPEC STATISTICAL BULLETIN.

_____. OPEC SPECIAL FUND. Vienna, Austria: OPEC, 1976-- . Annual.

Gives details of OPEC aid to underdeveloped countries.

Petro Consultant. FOREIGN SCOUTING SERVICE. Geneva: 1976-- . Monthly.

Data on oil exploration activity in 130 countries, including concessions, wildcat drilling, development drilling, production and reserves, rig activity, wells, crews, and active rigs.

PETROLEUM ECONOMIST. London: Petroleum Press Bureau, 1933-- . Monthly.

News and commentary on all aspects of the world oil and natural gas business. Covers prices, taxes, exports, exploration and discoveries, production, development, use and conservation, government policies relative to petroleum, offshore rigs, tankers, and related topics and their financial and economic implications. Also includes corporate news briefs and tables of product and security prices, production, and refinery output. Also available in French, German, Spanish, Arabic, and Japanese.

PETROLEUM ENGINEER INTERNATIONAL. Dallas: Petroleum Engineer Publishing Co., 1929-- . Quarterly.

Frequent statistical information on the petroleum industry.

Petroleum Equipment Suppliers Association. SERVICE POINT DIRECTORY 75/
76. Houston: 1975-- . Annual.

Contains membership listing of the Petroleum Equipment Suppliers
Association: key executive officers, geographical location, and
type of operation for each of the member companies. Lists more
than three thousand domestic locations from which the petroleum
industry is served. In addition, another 450 international locations are also listed.

Petroleum Industry Research Foundation. OUTLOOK FOR OIL TO 1990 AND
AFTER: OVERVIEW AND FINDINGS. New York: 1978.

_____. U.S. OIL SUPPLY AND DEMAND TO 1990. New York: 1977.

PETROLEUM OUTLOOK. Greenwich, Conn.: John S. Herold, 1948-- .
Monthly.

News and commentary on important trends and developments in
the petroleum industry, with particular attention to the financial
results and stock market activity of individual companies. Covers
the oil and gas industry, mainly in Canada and the United States.

"Petroleum 2000." OIL AND GAS JOURNAL (Tulsa, Okla.) 75 (August 1977):
entire issue.

Comprehensive statistical analysis of the U.S. petroleum industry
through 2000, in areas like exploration, production, transportation, refining, natural gas, petroleum chemicals, synthetic fuels,
and refining.

PIPELINE DIGEST. Houston: Universal News, 1963-- . Semimonthly.

Data on pipeline construction industry, including latest projects
planned, contracts, location, and contractor.

PLATT'S OILGRAM NEWS SERVICE. New York: McGraw-Hill, 1934-- .
Daily.

Newspaper concerned with international petroleum and energy
affairs includes other fuels. Covers financial and industrial aspects of energy industry.

PLATT'S OILGRAM PRICE SERVICE. New York: McGraw-Hill, 1923-- .
Loose-leaf. Daily.

Tables of prices for various petroleum products around the world, as well as spot tanker rates and market conditions in the national and regional markets in the United States.

REFINING, CONSTRUCTION, PETROCHEMICAL AND NATURAL GAS PROCESSING PLANTS OF THE WORLD. Tulsa, Okla.: Midwest Oil Register, 1976-- . Annual.

Data on worldwide refineries by type, capacity, and products manufactured.

Sales and Marketing Management. SURVEY OF BUYING POWER. New York: 1918-- . Annual.

Data on gasoline service station sales in leading metropolitan county areas. Now available also in the form of an expanded SMM data service.

Shriver Associates. U.S. OIL AND NATURAL GAS FINDINGS COSTS. 2 vols. Parsippany, N.J.: 1977.

Society of Exploration Geophysicists. GEOPHYSICAL ACTIVITY. Tulsa, Okla.: 1976-- . Annual.

Data on exploratory activity in the United States.

Statistical Office of the European Communities. PETROLEUM STATISTICS 1977. Luxembourg: 1979. 64 p.

Contains balance sheets for crude oil and petroleum products for 1976 and 1977 for European Community countries.

Sun Oil Co. ANALYSIS OF WORLD TANK SHIP FLEET. St. Davids, Pa.: 1945-- . Annual.

Contains data on oil transportation, in terms of U.S. tank ship fleet, world tank ship fleet, world tanker carrying capacity by size of vessel, world tank ship fleet by flag and year of construction, and world tank ships under construction or on order.

United Nations. Department of Economic and Social Affairs. PETROLEUM IN THE 1970'S. New York: 1974.

Report of the Ad-hoc Panel of Experts on the projections of demand and supply of crude petroleum and products. It contains projected inland consumption of petroleum products in developing countries; energy consumption in Latin America, Africa, and Asia; the centrally planned economies; production, consumption, and exports and imports of petroleum; capital and exploration expenditures in the petroleum industry by type and region; price structure; and forecasts.

U.S. Army. Corps of Engineers. WATERBORNE COMMERCE OF THE UNITED STATES. Washington, D.C.: Government Printing Office, 1953-- . Annual.

Data on domestic barge shipments of crude oil and petroleum products on the waterways of the United States.

U.S. Bureau of Mines. MINERAL INDUSTRY SURVEYS: AVAILABILITY OF HEAVY FUEL OILS BY SULFUR CONTENT. Washington, D.C.: Government Printing Office, 1976.

Contains data on stocks of residual and number 4 fuel oil held by refineries and bulk terminals by sulfur content by refinery and PAD district, imports of residual fuel oil by percent sulfur content by PAD district and country of origin and by state, and interdistrict waterborne movements of residual and fuel oils by sulfur content.

_____. MINERAL INDUSTRY SURVEYS: CARBON BLACK. Washington, D.C.: 1976.

Production shipments and value of carbon black in the United States.

_____. MINERAL INDUSTRY SURVEYS: LIQUEFIED PETROLEUM GAS SHIP-MENTS. Washington, D.C.: 1976.

_____. MINERAL INDUSTRY SURVEYS: MOTOR GASOLINES, AVERAGE OCTANE RATINGS. Washington, D.C.: 1976.

_____. MINERAL INDUSTRY SURVEYS: PETROLEUM STATEMENT. Washington, D.C.: Government Printing Office, 1976.

These are statements providing petroleum statistics on price, production, input, stocks, tanker movements, and consumption.

_____. MINERAL INDUSTRY SURVEYS: SHIPMENTS OF ASPHALT. Washington, D.C.: 1976.

_____. MINERAL INDUSTRY SURVEYS: SHIPMENTS OF FUEL OIL AND KEROSENE. Washington, D.C.: 1976.

_____. PETROLEUM REFINERIES IN THE UNITED STATES AND PUERTO RICO. Washington, D.C.: 1974-- . Annual.

U.S. Bureau of the Census. ANNUAL SURVEY OF OIL AND GAS. Washington, D.C.: Government Printing Office, 1973-- .

> Report on operating revenues and expenses of domestic oil and gas field exploration, development, and production, based on a survey of establishments cited in the Census of Mineral Industries.

_____. BUNKER OIL AND COAL LADEN IN THE UNITED STATES ON VESSELS ENGAGED IN FOREIGN TRADE. Washington, D.C.: Government Printing Office, 1951-- . Annual.

_____. CENSUS OF AGRICULTURE. Washington, D.C.: Government Printing Office, 1840-- . Quinquennial.

> Data on petroleum expenditures of farm business by state and type of product.

U.S. Central Intelligence Agency. CHINA: OIL PRODUCTION PROSPECTS. Washington, D.C.: Government Printing Office, 1977.

_____. EXPORT REFINING CENTERS OF THE WORLD. Washington, D.C.: Government Printing Office, 1975.

> Describes oil refinery capacity and export volume of five major non-Communist oil refinery areas: Caribbean area, Rotterdam, Persian Gulf, Italy, and Singapore. It gives the location, refining capacity, ownership of refineries, and exports by country.

_____. WORLD OIL REFINERIES. Washington, D.C.: Government Printing Office, 1974-- . Annual.

> A country-by-country account of world crude oil refinery capacity along with the following additional data: ownership of world oil refinery capacity--private as well as government, area refining capacity by refinery and ownership, company by area, and oil companies' control of production in OPEC countries.

U.S. Congress. Senate. Interior and Insular Affairs Committee. TRANS-ALASKA PIPELINE AND WEST COAST PETROLEUM SUPPLY, 1977-82. 93d Cong., 2d sess. Washington, D.C., 1974.

> Contains postembargo projections of probable volumes and destinations of Alaskan north slope crude oil 1977-82, and marketing plans of major oil companies and the federal government over the first five years of trans-Alaska pipeline operations.

U.S. Congress. Senate. Judiciary Committee. Subcommittee on Antitrust and Monopoly. PETROLEUM INDUSTRY. 94th Cong., 1st. sess. Washington,

D.C.: Government Printing Office, 1975.

Contains data on vertical integration in the petroleum industry in terms of production, refining, pipeline, and marketing. Includes information on refining capacity, financial data and rates of return, share ownership of the six largest U.S. oil companies, and demand and sales of gasoline and distillates.

U.S. Congress. Senate. Senate Committee on Small Business. INTERNATIONAL PETROLEUM CARTEL. 93d Cong., 2d sess. Washington, D.C.: Government Printing Office, 1975.

Reprint of the 1952 Federal Trade Commission report containing historical data on resources and concentration in the world petroleum industry, trade patterns, and company joint ownerships and directorates.

U.S. Department of the Interior. Office of Oil and Gas. PETROLEUM SUPPLY AND DEMAND IN THE NON-COMMUNIST WORLD. Washington, D.C.: Government Printing Office, 1971-- . Annual.

An inventory of crude oil production by source, refined product output by source, supply and demand of crude oil and products, crude oil production and imports, international flow of petroleum, estimated availability and utilization of tankers, refining yields by country or geographical area, and estimated proven crude oil resources in the non-Communist world.

_____. WORLD WIDE CRUDE OIL PRICES. Washington, D.C.: 1976-- . Annual.

Ready reference on foreign crude oil prices. It contains data on total posted price, landed cost charter, delivered price, U.S. and landed spot price, harbor dues, royalty, average producing cost, average government take, and summaries of various oil agreements currently in force.

U.S. Department of the Interior. Office of the Secretary. UNITED STATES PETROLEUM THROUGH 1980. Washington, D.C.: Government Printing Office, 1968.

Current and projected data on U.S. supply and demand for petroleum, crude oil production, and productive capacity of natural gas in the United States, cost of drilling and equipping oil and gas wells, depth of producing wells in the United States, and research and development expenditures.

U.S. Energy Information Administration. MONTHLY PETROLEUM PRODUCT PRICE REPORT. Springfield, Va.: National Technical Information Service, 1978-- .

———. MONTHLY PETROLEUM STATISTICS REPORT. Springfield, Va.: National Technical Information Service, 1978-- .

———. MOTOR GASOLINE SUPPLY AND DEMAND, 1967-78. Washington, D.C.: Government Printing Office, 1977-- . Annual.

> Reports on domestic motor gasoline production, consumption, trade and stocks, including historical trend data from 1967-78.

U.S. Federal Energy Administration. INITIAL REPORT ON OIL AND GAS RESOURCE, RESERVES, AND PRODUCTIVE CAPACITIES. Washington, D.C.: 1975

> Contains an analysis of U.S. oil and gas resources, reserves, and petroleum refining capacity through 1985.

———. NATIONAL PETROLEUM PRODUCT SUPPLY AND DEMAND. Publication no. PB 254 969. Springfield, Va.: National Technical Information Service, 1976. Annual.

> Documentation of the short-term petroleum product supply and demand through 1978, assuming a composite low demand, a composite high demand, and a low domestic supply.

———. THE PETROLEUM INDUSTRY: A REPORT ON CORPORATE AND INDUSTRY STRUCTURE AND OWNERSHIP. Washington, D.C.: Government Printing Office, 1975.

> This report is a kind of who's who in oil. It is the most up-to-date compendium of raw data describing ownership in petroleum trade. This 850-page document lists parent companies; energy-related applications: coal, natural gas, geothermal energy, and uranium firms. The report is available from the National Technical Information Service, Springfield, Virginia.

———. PETROLEUM SUPPLY AND DEMAND IN THE NON-COMMUNIST WORLD. Washington, D.C.: Government Printing Office, 1972-- . Annual.

> Contains data on petroleum supply, demand, production, and transport data for all countries of the free world.

———. TRENDS IN REFINERY CAPACITY AND UTILIZATION. PETROLEUM REFINERIES IN THE UNITED STATES. Washington, D.C.: Government Printing Office, 1974-- . Annual.

> Report on U.S. refining capacity from 1960 to 1980. It contains data on actual crude runs, refinery expansion, construction and reactivation, world refining exporting centers, and the refinery and net exportable capacities of centers in the Caribbean area, Middle East, Canada, Italy, and Singapore.

_____. WEEKLY DEMAND WATCH. Washington, D.C.: Government Printing Office, 1976-- .

This is a news release which contains data on the total energy demand, demands for motor gasoline, residual, distillate and other products.

U.S. Federal Energy Administration. National Energy Information Center. MONTHLY SALES OF PETROLEUM REFINED PRODUCTS. Washington, D.C.: Government Printing Office, 1976-- .

Measures changes in marketing patterns and aggregate market share of petroleum refinery products, distillate fuel oil, kerosene, jet fuel, residual fuel, propane, and aviation gasoline.

_____. PETROLEUM MARKET SHARES. Washington, D.C.: Government Printing Office, 1976-- . Monthly.

Monitors changes in the aggregate market shares of motor gasoline dealers. Contains, among other things: market share and average price by brand category, gallons sold through service stations by market type, number of gasoline service stations, and average sales by marketer.

U.S. Federal Energy Administration. Office of Energy Statistics. MONTHLY PETROLEUM STATISTICS REPORT. Springfield, Va.: National Technical Information Service, 1975-- .

Data on production, imports, and stocks of crude oil, motor gasoline, jet fuels, distillate fuel oil and residual fuel oil; refinery operations; and prices of petroleum and related products.

_____. THE PIMS U.S.-OPEC PETROLEUM REPORT. Publication no. PB 245 307/4. Springfield, Va.: National Technical Information Service, 1973.

Data compiled by the Petroleum Industry Monitoring System. Deals with statistics on petroleum imports from OPEC countries. Includes petroleum products and prices.

U.S. Geological Survey. GEOLOGICAL ESTIMATES OF UNDISCOVERED RECOVERABLE OIL AND GAS RESOURCES IN THE UNITED STATES. Geological Survey Circular, no. 725. Washington, D.C., 1975.

Estimates offshore and onshore cumulative production, reserves, and undiscovered recoverable resources of crude oil, natural gas, and natural gas liquids, in fifteen production regions including Alaska.

_____. LATIN AMERICA'S PETROLEUM PROSPECTS IN THE ENERGY CRISIS. Geological Survey Bulletin, no. 1411. Washington, D.C.: Government Printing Office, 1975.

Analysis of Latin American energy supply patterns and the

outlook for development of its petroleum resources. It shows for each Latin American country: per capita GNP and estimated energy consumption, selected petroleum exchange balances, and estimated extent of petroleum resources, and exploratory drilling.

_____. MINERAL PRODUCTION, ROYALTY INCOME AND RELATED STATISTICS. Washington, D.C.: Government Printing Office, 1976-- .

Data on production of crude oil and natural gas on U.S. government lands, naval petroleum reserves, and Indian lands.

U.S. Geological Survey. Conservation Division. FEDERAL AND INDIAN LANDS OIL AND GAS PRODUCTION. ROYALTY INCOME AND RELATED STATISTICS. Washington, D.C.: Government Printing Office, 1971-- . Annual.

Report on the oil and gas leasing on- and offshore, well activity by state, production and revenue, value and royalty. The data on oil and condensate gas, natural gasoline and LPG are given for all lands, public lands, outer continental shelf, acquired lands, naval petroleum reserves and, Indian lands.

U.S. International Trade Commission. FACTORS AFFECTING WORLD PETROLEUM PRICES TO 1985. Washington, D.C.: Government Printing Office, 1977.

_____. SYNTHETIC ORGANIC CHEMICALS. Washington, D.C.: 1917-- . Annual.

Production and sales of crude products from crude oil and natural gas for chemical conversion in the United States.

_____. WORLD OIL SUPPLY AND DEMAND TO 1990. Washington, D.C.: Government Printing Office, 1977.

U.S. Office of Technology Assessment. ENHANCED OIL RECOVERY POTENTIAL IN THE UNITED STATES. Washington, D.C.: Government Printing Office, 1978.

U.S. OIL WEEK. Arlington, Va.: Observer Publishing Co., 1964-- . Weekly.

Commentary on petroleum product dealers and marketers. Statistics on oil company activities and petroleum product supply and demand.

Utah. Oil and Gas Conservation. YEARLY WELL COMPLETION REPORT: WELLS COMPLETED OR ABANDONED. Salt Lake City: 1976-- . Annual.

WALTER SKINNER'S OIL AND GAS INTERNATIONAL YEARBOOK, 1976-77. London: Financial Times, 1976.

WORLD OIL. Houston: Gulf Publishing Co., 1916-- . Monthly.

Frequent statistical information on petroleum industry.

WORLD PETROLEUM REPORT. New York: M. Palmer Publishing Co., 1957-- . Annual.

Annual review of international oil operations, containing tabulations of world petroleum data by country including the following: crude oil production in barrels per day, proven and provable crude oil reserves in million barrels, consumption and percentage change from previous year, number of refineries, crude capacity and major downstream processing capacities with short run estimates for three years, and a review of the petroleum industry by country.

WORLDWIDE DIRECTORY OF PIPELINES AND CONTRACTORS. Tulsa, Okla.: Petroleum Publishing Co., 1977-- .

Data on more than three hundred operating companies and six hundred construction engineering-service firms. Includes important facts. Tables and charts show miles of line, size, compressor and pump stations, deliveries, investment, operating revenues, net income, and a current survey of lines under construction.

WORLDWIDE PETROCHEMICAL DIRECTORY. Tulsa, Okla.: Petroleum Publishing Co., 1977-- .

Data on more than 15,000 personnel, 1,200 companies, 2,400 locations. Includes surveys of plants, giving feed stock, principal products, and plants under construction.

Wyoming. Oil and Gas Conservation Commission. WYOMING OIL AND GAS STATISTICS. Casper: 1976-- . Annual.

SOLAR AND OTHER EMERGING SOURCES

Bacher, Ken. PUTTING THE SUN TO WORK. Phoenix: Arizona Fuel and Energy Office, 1974.

This is a history and directory of currently available solar energy applications.

Bennington, G. SOLAR ENERGY: A COMPARATIVE ANAYSIS TO THE YEAR 2020. McLean, Va.: Mitre Corp., 1978.

Berman, Edward R. GEOTHERMAL ENERGY. Energy Technology Review Series. Parkridge, N.J.: Noyes Data Corp., 1975.

Deals with historical background and recent explorations of geo- thermal energy and U.S. agencies involved. It contains a world survey of major geothermal installations. The cost factor is also given careful consideration.

Colorado. Geological Survey. GEOTHERMAL RESOURCES OF COLORADO. By Richard Howard Pearl. Denver, 1972.

Environment Information Center. SOLAR UPDATE. New York: 1977.

A special compendium of the EIC ENERGY DIRECTORY UPDATE devoted exclusively to solar energy.

Hagen, Arthur W. THERMAL ENERGY FROM SEA. Energy Technology Review Series. Parkridge, N.J.: Noyes Data Corp., 1975.

The ocean is a reservoir of solar energy and allows sea thermal power systems to operate all the year round. The book aims to provide a data base to facilitate experience to prove the econom- ic feasibility of such systems.

INFORMAL DIRECTORY OF THE ORGANIZATIONS AND PEOPLE INVOLVED IN THE SOLAR HEATING OF BUILDINGS. Cambridge, Mass.: E.A. Shur- cliffs, 1975.

Jackson, Frederick R. ENERGY FROM SOLID WASTE. Energy Technology Review Series. Parkridge, N.J.: Noyes Data Corp., 1974.

Based on information collected through the U.S. Environmental Protection Agency experiments on the practicability of energy from solid waste. It gives data on the various projects for low sulfur fuel--like the St. Louis project, and Philadelphia and Cleveland feasibility studies. It examines the current European practices. Detailed data on process description are provided.

Martz, C.W., ed. SOLAR ENERGY SOURCE BOOK. Washington, D.C.: Solar Energy Institute of America, 1977.

Patton, Arthur R. SOLAR ENERGY FOR HEATING AND COOLING OF BUILDINGS. Energy Technology Review Series. Parkridge, N.J.: Noyes Data Corp., 1975.

Describes several large-scale feasibility studies with designs suit- able for institutions and industrial plants. The federally spon- sored studies, like the Westinghouse, General Electric, and TRW studies, are dealt with in detail.

Simmons, Daniel M. WIND POWER. Energy Technology Review Series. Parkridge, N.J.: Noyes Data Corp., 1975.

> Deals with wind behavior, site selection, wind machine design, conversion and storage systems, economic feasibility studies, and commercially available wind machines.

THE SOLAR DATA DIRECTORY OF THE SOLAR ENERGY INDUSTRY. Hampton, N.H.: Solar Data, 1975.

SOLAR DIRECTORY. Denver: Environmental Action of Colorado, University of Colorado, 1975.

SOLAR ENERGY AND RESEARCH DIRECTORY. Ann Arbor, Mich.: Ann Arbor Science Publishers, 1976-- . Annual.

> Data on manufacturers, design and construction firms, researchers, government-sponsored research and development groups, conservationists, and distributors.

SOLAR ENERGY DIRECTORY: A DIRECTORY OF DOMESTIC AND INTERNATIONAL FIRMS INVOLVED IN SOLAR ENERGY. Phoenix, Ariz.: Centerline Corp., 1976-- . Annual.

> Data classified by occupation like architects, associations, societies, contractors, consultants, engineers, government agencies, manufacturers, market distribution, and research development. Includes some foreign listings.

SOLAR ENERGY INDUSTRY DIRECTORY AND BUYERS GUIDE. Washington, D.C.: Solar Energy Industry Association, [1975?].

U.S. Congress. House. Committee on Science and Technology. Subcommittee on Energy Research and Development and Administration. SOLAR HEATING AND COOLING DEMONSTRATION ACT OF 1974. OVERSIGHT HEARINGS. 93d Cong., 2d sess. Washington, D.C.: Government Printing Office.

U.S. Department of Energy. SOLAR ENERGY, A STATUS REPORT. DOE/ET-0062. Washington, D.C.: 1978.

U.S. Energy Research and Development Administration. ECONOMIC ANALYSIS OF SOLAR WATER AND SPACE HEATING. Washington, D.C.: Government Printing Office, 1976.

> Examines the competitive status of solar heating and hot water systems for new single-family residences, compared to conventional systems using natural gas, fuel oil, electric resistance, or electric heat pumps. Cost comparisons are made for thirteen

cities based on local climatic conditions and cost alternatives us-
ing solar collectors costing $20.00, $15.00, or $10.00 per square
foot.

_____. FEDERAL WIND ENERGY PROGRAM. Springfield, Va.: National
Technical Information Service, 1975.

Contains abstracts of National Science Foundation and ERDA-
funded research and development projects related to wind energy
system costs, reliability, technical feasibility, and social and
environmental impacts.

_____. NATIONAL PLAN FOR SOLAR HEATING AND COOLING. Research
and Commercial Applications series. Washington, D.C.: Government Printing
Office, 1975.

Contains tables projecting the number of houses and commercial
buildings with installed solar energy systems, the resulting
annual fuel savings (1975-85), and the funding estimates.

_____. NATIONAL PROGRAM FOR SOLAR HEATING AND COOLING OF
BUILDINGS. Washington, D.C.: Government Printing Office, 1976-- .
Annual.

Review of the status of federal research and development and
demonstration projects for promotion of solar heating and cooling
in commercial and residential buildings. Includes lists of residential and commercial demonstration projects, showing location,
contractor, building type, and solar energy application.

_____. NATIONAL SOLAR ENERGY RESEARCH AND DEVELOPMENT AND
DEMONSTRATION PROGRAM DEFINITION REPORT. Washington, D.C.:
Government Printing Office, 1975.

Report defining national solar energy development goals and
implementation plans to 1985. It discusses the technology and
applications. Estimates total energy to be supplied through this
source and the cost.

_____. PACIFIC REGIONAL SOLAR HEATING HANDBOOK. 2d ed. Wash-
ington, D.C.: Government Printing Office, 1977.

Contains climatic and related data on California, Oregon, Washington, and Arizona, useful in designing a solar heating system
for a residential or small commercial buildings in the Pacific
Coast region.

U.S. Energy Research and Development Administration. Division of Solar
Energy. SOLAR ENERGY HEATING AND COOLING PRODUCTS. Washington, D.C.: 1975.

U.S. Federal Energy Administration. BUYING SOLAR. By Joe Dawson. Washington, D.C.: Government Printing Office, 1976.

Guide to the purchase of solar heating equipment in terms of available systems, system operations, and methods for calculating annual fuel savings for solar-fitted homes. Includes interspersed tables on heating degree days and solar radiation values for major U.S. cities; solar system costs; and estimated payback periods and savings under various energy source, equipment, geographic, and architectural alternatives.

_____. PROJECT INDEPENDENCE TASK FORCE FINAL REPORT ON SOLAR ENERGY. Washington, D.C.: Government Printing Office, 1975.

Forecasts viable solar energy conversion systems before the year 2000. Six technically feasible solar energy technologies for heating and cooling of building providing high temperature heat and producing electric power are described. Projects annual energy contribution by these sources and estimates equivalent conventional fuel savings.

_____. SOLAR COLLECTOR MANUFACTURING ACTIVITY. Washington, D.C.: Government Printing Office, 1976-- . Semiannual.

Survey of production and sales of medium- and low-temperature solar collectors by public and private firms. Details include rate, square feet, list of manufacturing collectors.

_____. SOLAR ENERGY PROJECTS OF THE FEDERAL GOVERNMENT. Springfield, Va.: National Technical Information Service, 1975. Publication no. PB 241 620.

Report on projects of fourteen federal agencies to develop or promote solar energy applications. Covering heating and cooling of buildings, conversion of ocean heat and wind to electricity, photovoltaics, and bioconversion of waste materials to fuel.

U.S. Geological Survey. ASSESSMENT OF GEOTHERMAL RESOURCES OF THE U.S., 1975. Geological Survey Circular, no. 726. Washington, D.C.: 1975.

Compilation of data on the magnitude, distribution, and recoverability of U.S. geothermal resources in 1975, by major type.

U.S. Library of Congress. Congressional Research Service. Science Policy Research Division. SURVEY OF SOLAR ENERGY PRODUCTS AND SERVICES. Washington, D.C.: Government Printing Office, 1975.

Contains a description of solar energy products and services, such as solar collectors, collectors designed for swimming pools, total solar building systems, control and cooling systems. Also gives directory-type information on people involved in research, applications, and professional societies.

COMPOSITE SOURCES

Unlike the parts covered so far in section III, this part is devoted to a selectively annotated listing of sources which provide statistical data, not on just one type of energy but on all types, including coal, electricity, natural gas, nuclear power, petroleum, and solar. Some of the sources belong to the factbook or almanac type, containing information on all general topics including energy.

Alabama. University. Center for Business and Economic Research. ECO-NOMIC ABSTRACT OF ALABAMA. University: 1975.

Alaska. Department of Commerce and Economic Development. Division of Economic Enterprise. ALASKA STATISTICAL REVIEW. Juneau: 1972.

Allen, Edward et al. "U.S. Energy and Economic Growth 1975-2000." Oak Ridge, Tenn.: Institute for Energy Analysis, 1976.

Arizona. University. Division of Economic and Business Research. ENERGY CONSUMPTION BY STATE HISTORICAL TRENDS FOR 1957-1971 AND PRO-JECTIONS TO 1975 AND 1980. By Helmut J. Frank and Jean E. Weber. Tucson, 1973.

Arkansas Almanac, Inc. ARKANSAS ALMANAC. Little Rock. Annual.

Arnold Bernhard & Co. THE VALUE LINE. New York: 1937-- . Loose-leaf.
Investment advisory service containing financial and related data on all energy companies.

Bankers Trust Co. CAPITAL RESOURCES FOR ENERGY THROUGH THE YEAR 1990. New York: 1976.

Battelle Memorial Institute. ENERGY INFORMATION RESOURCES. Compiled by Patricia L. Brown et al. Sponsored by National Science Foundation. Columbus, Ohio: Columbus Laboratories; Washington, D.C.: American Society of Information Science, 1975.
This is an inventory of energy research and development information on the power resources of the continental United States, Hawaii, and Alaska.

Behling, D.J. ANALYSIS OF PAST AND EXPECTED FUTURE TRENDS IN THE

U.S. ENERGY CONSUMPTION, 1947-2000. Upton, N.Y.: Brookhaven National Laboratory, 1977.

Broadman, John R. A COMPARISON OF ENERGY PROJECTIONS TO 1985. Paris: International Atomic Energy Agency, 1979.

Building Owners and Managers Association. ANNUAL OFFICE BUILDING EXPERIENCE EXCHANGE REPORT. Chicago: 1920.

> Analysis of operating costs include expenses in electrical systems, heating, and air conditioning.

Bullard, Clark; Hannon, Bruce; and Herendeen, Robert. ENERGY FLOW THROUGH THE UNITED STATES ECONOMY: A WALL CHART. Urbana: University of Illinois Press, 1977.

> Data on energy use and flow patterns in the United States.

Business Publishers, Inc. ENERGY RESOURCES REPORT. Silver Spring, Md.: 1973-- . Weekly.

> Information on world energy supply trends, exploratory and other programs, and corporations engaged in the energy fields.

California. Department of Finance. CALIFORNIA STATISTICAL ABSTRACT. Sacramento: 1960-- . Annual.

CAPITAL ENERGY LETTER. Washington, D.C.: Capital Energy Letter, Inc., 1974-- . Loose-leaf. Weekly.

> Contains news and reports of federal legislation, hearings and agency decisions and policies relating to coal, electricity, natural gas, petroleum, and nuclear energy.

Charter House Group. JANE'S MAJOR COMPANIES OF EUROPE. London: 1965-- . Annual.

> Directory providing information on major European coal, oil, gas, and mining companies. The data include management, subsidiaries, products, factories and plants, employment, capitalization, and long-term debt. The countries covered are Austria, Belgium, Denmark, Finland, France, West Germany, Republic of Ireland, Italy, Luxembourg, Netherlands, Norway, Portugal, Sweden, Switzerland, and the United Kingdom.

Commission on European Communities. THE OECD AREA'S NET ENERGY IMPORTS TO 1985. Brussels: 1976.

Commodity Research Bureau, Inc. COMMODITY YEARBOOK. New York.

The energy related commodities covered are coal and coke, electric power, petroleum, gas, and uranium. These are analyzed from the point of view of production, consumption and utilization, stocks, imports and exports, and market price structure.

_____. STATISTICAL ABSTRACT SERVICE. New York: 1929-- . Monthly.

Contains updating price and related data on energy commodities contained in the COMMODITY YEARBOOK (see above).

Conference Board. ENERGY CONSUMPTION IN MANUFACTURING. New York: 1975.

A report submitted under the Energy Policy Project of the Ford Foundation. It projects energy use in manufacturing to 1980. The following industry groups are studied: food; paper; chemical, petroleum, and coal products; stone, clay, and glass industries; and primary industries. These account for four-fifths of energy used in manufacturing.

Connecticut. Department of Commerce. Technical Services Division. CONNECTICUT MARKET DATA. Hartford: 1959-- . Annual.

CONSUMER EUROPE. London: Euromonitor Publications, 1976-- . Annual.

Comprehensive handbook on consumer markets in Europe containing systematic marketing profiles for over one hundred consumer products in eleven major countries, including Austria, Belgium, Denmark, France, Germany, Great Britain, Italy, Netherlands, Spain, Sweden, and Switzerland.

Corpus Publishers, Ltd. ENERGY ANALECTS. Toronto: 1972-- . Weekly.

Covers North American energy industries. Contains, among other things, statistics on reserves, production, fuel prices, marketing, and environmental impact.

Dallas Morning News. TEXAS ALMANAC, 1976-1977. Dallas: 1975.

Darmstadter, J. ENERGY IN THE WORLD ECONOMY. Published for the Resource for Future. Baltimore: Johns Hopkins Press, 1971.

A statistical review of the trends in output, trade, and consumption since 1925, dealing with the quantitative aspects of long-term trends in energy consumption, production, and foreign trade. Discusses the change during this century of the world's fuel base away from coal towards oil and natural gas.

_____. HOW INDUSTRIAL SOCIETIES USE ENERGY: A COMPARATIVE
ANALYSIS. Baltimore: Johns Hopkins University, 1977.

Delaware. State Planning Office. DELAWARE STATISTICAL ABSTRACT.
Dover: 1975.

Doernberg, A. ENERGY USE IN JAPAN AND THE UNITED STATES. Upton,
N.Y.: Brookhaven National Laboratory, 1977.

Dunkerley, Joy. INTERNATIONAL COMPARISON OF ENERGY CONSUMP-
TION. Washington, D.C.: Resources for the Future, 1977.

Emmings, Steven D. MINNESOTA: HISTORICAL DATA ON FUEL AND
ELECTRICITY. Minneapolis: Minnesota Energy Project, State Planning Agency,
1974.

ENERGY: GLOBAL PROSPECTS 1985-2000. By the Workshop on Alternative
Energy Strategies. New York: McGraw-Hill, 1977.

> Data on global supply and demand for energy, and forecasts
> through the year 2000.

ENERGY DEMAND STUDIES: MAJOR CONSUMING COUNTRIES. ANALYSIS
OF 1972 DEMAND AND PROJECTIONS OF 1985 DEMAND. Cambridge:
MIT Press, 1976.

> This is the first technical report of the Workshop on Alternative
> Energy Strategies (WAES). Contains accurate and comparable
> estimates of future energy demand. Data cover Canada, Den-
> mark, Finland, France, West Germany, Iran, Italy, Japan,
> Mexico, the Netherlands, Norway, Sweden, the United Kingdom,
> and the United States.

ENERGY FACT BOOK 1976. Arlington, Va.: Tetra Tech, 1975.

> Contains statistics on various kinds of energy resources.

ENERGY HISTORY OF THE UNITED STATES, 1776-1976. WALL CHART.
Washington, D.C.: U.S. Energy Research and Development Administration.
Available from the Government Printing Office.

> ERDA's wall chart and the accompanying twenty-four page pam-
> phlet explain the impact of energy on the nation's economy.

ENERGY LAW AND REGULATIONS. Proceedings of the Energy Law Seminar,
January 27-29, 1976. Washington, D.C.: Government Institutes, 1976.

> Provides a wide range of legal and related information including
> a forecast of world energy supplies and demand.

ENERGY STATISTICS. Brussels, Belgium: Statistical Office of the European Communities, 1976-- . Quarterly.

Provides data on the overall energy balance sheet concerning major items of energy supplies and disposals for the whole European Economic Community and for each member country. It also gives the balance sheet and main monthly statistical series available for each source of energy. The most recent annual data and updated figures are shown for the balance sheets as well as for the various statistical series.

Environment Information Center. ENERGY DIRECTORY AND UPDATE. New York: 1973-- . Annual.

Comprehensive guide to the nation's energy organizations, decision makers, and information sources. Details include description of programs and policies and amount of dollars spent in the area of energy.

_____. ENERGY INDEX. New York: 1975-- . Annual.

A statistical, bibliographic, and directory-type publication dealing with energy resources and reserves, energy consumption, conversion factors, financial profiles of energy companies, and energy-related environmental impact. All types of energy are dealt with specifically oil and gas, coal, oil shale, nuclear energy, hydroelectric power, and unconventional energy sources like solar, geothermal, and organic waste.

_____. ENERGY INFORMATION LOCATOR. New York: 1976-- . Annual.

Selective guide to information centers, systems, data bases, abstracting services, directories, newsletters, binder services, and journals in the area of energy. Includes complete bibliographic data like profiles, prices, names, addresses, and telephone numbers.

Exxon Corporation. Public Affairs Dept. WORLD ENERGY OUTLOOK. New York: 1978.

Finn, M.G. ENERGY-RELATED SCIENTISTS AND ENGINEERS: STATISTICAL PROFILE. Oak Ridge, Tenn.: Oak Ridge Associated Universities, 1978.

Folk, H. SCIENTIFIC AND TECHNICAL PERSONNEL IN ENERGY-RELATED ACTIVITIES: CURRENT SITUATION AND FUTURE REQUIREMENTS. Urbana: University of Illinois, Center for Advanced Computation, 1977.

Florida. University of. Bureau of Business and Economic Research. FLORIDA

STATISTICAL ABSTRACT. Gainsville: 1967-- . Annual.

Foster Associates. ENERGY PRICES, 1960-1973. Cambridge, Mass.: Ballinger
Publishing Co., 1974.

> This was one of the reports submitted under the Ford Foundation
> Energy Policy Project. It contains a compilation of information
> on energy prices, that is, retail and wholesale prices for primary
> and secondary energy sources in the United States.

Franssen, Herman T. ENERGY AN UNCERTAIN FUTURE. Senate Committee
Print 95-157. Washington, D.C.: Government Printing Office, 1978. 329 p.

> Presents all energy supply/demand forecast for the United States
> and world through the year 1990.

Fremond, Felix. WORLD MARKETS OF TOMORROW. New York: Harper,
1972.

> Contains world growth projections in the energy field as follows:
> use of energy, including electricity, hydro, and nuclear,
> through 2000; contributions of primary sources in meeting world
> energy needs; total use of energy; per capita use of energy by
> country; installed electric capacity in kilowatts; and U.S. kilo-
> watt per capita and projections.

Georgia. University of. College of Business Administration. Division of Research. GEORGIA STATISTICAL ABSTRACT. Athens: 1951-- . Annual.

Good, Barry C. WORLD ENERGY BALANCES: THE NEXT FIFTEEN YEARS.
New York: Morgan Stanley & Co., 1976.

Gordian Associates. ENERGY CONSERVATION, THE DATA BASE: THE
POTENTIAL FOR ENERGY CONSERVATION IN NINE SELECTED INDUSTRIES.
9 vols. Conservation Paper Series, no. 9-17. Washington, D.C.: Office
of Energy Conservation and Environment, Office of Industrial Programs, 1975.
Available from the Government Printing Office.

> This report contains statistical data on energy consumption and
> conservation in the United States by selected industry, including
> plastics, petroleum refining, cement, aluminum, steel, glass,
> paper products, land, styrene, and rubber.

Government Institutes, Inc. ENERGY REFERENCE HANDBOOK: A GLOSSARY
WITH ABBREVIATIONS AND CONVERSION TABLES. Washington, D.C.: 1974.

> This is a convenient reference book containing the terminology
> and conversion tables including conversion tables for temperature,
> length, area, volume, force, density, pressure, rate of thermal

conductance, heat flow, viscosity, kinematic viscosity, angular velocity, and thermal conductivity. It also has the English conversion table, metric conversion table, numerical abbreviations, periodic table of the elements, and others.

Great Britain. Central Office of Information. Reference Division. ENERGY. British Industry Today Series. London: HMSO, 1975.

Provides information on the power resources in the United Kingdom.

Great Britain. Department of Energy. DIGEST OF THE UNITED KINGDOM ENERGY STATISTICS. London: HMSO, 1947-- . Annual.

Great Britain. Department of Energy. Economics and Statistics Division. ENERGY TRENDS. London: HMSO, 1976-- . Monthly.

Statistical bulletin containing data on the production, stocks, consumption, and utilization of coal, gas, electricity, and petroleum in the United Kingdom.

Gulf Publishing Co. HPI MARKET DATA. Houston: 1922-- . Annual.

This special annual market report on the hydrocarbon processing industry (HPI) examines HPI energy and construction trends, petrochemical, refining, gas processing, solid fuel processing, HPI equipment, design and maintanance trends, HPI capital spending, HPI market, and HPI buying power.

Gustaferro, Joseph F. FORECAST OF LIKELY U.S. ENERGY SUPPLY/DEMAND BALANCES FOR 1985 AND 2000. 2 vols. Springfield, Va.: National Technical Information Service, 1978. PB 266240; PB 287 487.

Hausgaard, Olaf. CONSUMPTION OF ENERGY IN NEW YORK STATE 1972 WITH ESTIMATES FOR 1973. Albany: New York State Public Service Commission, 1974.

Hawaii. Department of Planning and Economic Development. THE STATE OF HAWAII DATA BOOK 1976: A STATISTICAL ABSTRACT. 10th ed. 1976.

Hendrickson, Thomas A., ed. SYNTHETIC FUELS DATA BOOK. Denver, Colo.: Cameron Engineers, 1975.

Statistical and related data on all types of synthetic fuels.

Herman, Stewart; Cannon, James; and Inform Inc. ENERGY FUTURES: INDUSTRY AND THE NEW TECHNOLOGIES. Cambridge, Mass.: Ballinger, 1977.

Most comprehensive data on corporate activity in new energy technology areas produced to date of publication.

Hirst, Eric. RESIDENTIAL ENERGY USE TO THE YEAR 2000. Oak Ridge, Tenn.: Oak Ridge National Laboratory, 1977.

Estimates residential energy use, costs and savings potential under seven conservation options, through the years 1977-2000.

Idaho. University. Center for Business Development and Research. IDAHO STATISTICAL ABSTRACT. Moscow, Idaho: 1971.

Illinois. Department of Business and Economic Development. ILLINOIS INDUSTRIAL ENERGY CONSUMPTION DURING 1971 AND 1975. Springfield: 1976.

_____. ILLINOIS STATE AND REGIONAL ECONOMIC DATA BOOK. Springfield: 1970-- . Biennial.

Indiana. State Planning Services Agency. INDIANA FACT BOOK. Indianapolis, 1976.

INDUSTRIAL FUEL MARKETS, 1973-1974. Epping, Engl.: Gover Press, 1973.

Statistics on fuel trade, consumption, and prices in the United Kingdom.

International Research Group. ENERGY POLICIES IN THE EUROPEAN COMMUNITY. Washington, D.C.: Energy Research and Development Administration, 1975.

Traces the evolution of Europe's primary energy consumption structure. Contains projections of future energy consumption. Tables include actual internal consumption of primary energy, pre- and post-oil-crisis projections to 1985, and trends in energy consumption in the European Economic Community.

Ion, D.C. AVAILABILITY OF WORLD ENERGY RESOURCES. London: Graham and Trotman, 1975.

Provides statistical information on the following: world estimated ultimate crude oil recovery; world resources of uranium and thorium; energy content of proved resources; known coal reserves; world estimated crude oil recovery in discovered fields; Middle East resources; natural gas reserves by country; energy production by major region; synthetic fuel production; USSR energy production, demand of petroleum products, slurry and oil and gas pipelines, OPEC exports and revenues; forecasts of role of nuclear

energy in the world 1980-2000 and EEC forecasts of energy supply to 1985.

Iowa. Development Commission. Research Division. 1977 STATISTICAL PRO-
FILE OF IOWA. Des Moines: 1977.

Kansas. University. Institute for Social and Environmental Studies. KANSAS
STATISTICAL ABSTRACT. Lawrence: 1966-- . Annual.

Kentucky. Department of Commerce. KENTUCKY DESKBOOK OF ECONOMIC
STATISTICS. Frankfort: 1964-- . Annual.

Latin American Center. STATISTICAL ABSTRACT OF LATIN AMERICA. Los
Angeles: University of California, 1950-- . Annual.

> Data on the production of coal, petroleum, and natural gas,
> production of and trade in petroleum products, energy production
> and consumption, and electrical energy production and installed
> capacity in Latin American countries.

Liepins, G.D. BUILDING ENERGY USE DATA BOOK. Oak Ridge, Tenn.:
Oak Ridge National Laboratory, 1978.

McGraw-Hill, Inc. MINERAL RESOURCES INDUSTRIES CORPORATE PROFILES.
New York: Published jointly with Disclosure Inc., 1976.

> Comprehensive references on interindustry integration, capitaliza-
> tion and expansion plans of five hundred leading companies in the
> area of coal mining, coal reserves holders, petroleum producers,
> gas and transmission companies, metal and nonmetal mining com-
> panies, nonferrous smelting and refining firms, iron and steel
> producers, and top electric utilities.

Maine. State Development Office. FACTS ABOUT INDUSTRIAL MAINE.
Augusta: 1975-- . Updated continually.

Margolis Industrial Services. ENERGY REPORT. New York: 1974-- .
Monthly.

> Evaluation of technological and economic trends in petroleum,
> gas, coal, nuclear, geothermal, solar, shale, oil, and tar sands.

Maryland. Department of Economic and Community Development. MARYLAND
STATISTICAL ABSTRACT. Annapolis: 1977.

Massachusetts. Department of Commerce and Development. FACT BOOK.
Boston: 1976.

Metals Week. EIGHT MINERAL CARTELS. New York: 1975.

> Describes eight mineral cartels having significant effect on prices and supply. They control the following: oil, bauxite, copper, iron ore, tin, mercury, tungsten, and lead. The OPEC is analyzed thoroughly and there is also a profile of the oil industry.

Michigan. State University. Graduate School of Business Administration. Division of Research. MICHIGAN STATISTICAL ABSTRACTS. East Lansing: 1966-- . Annual.

Minnesota. Department of Economic Development. Research Division. MINNESOTA STATISTICAL PROFILE. St. Paul: 1975.

Minnesota. State Planning Agency. Office of Local and Urban Affairs. MINNESOTA POCKET DATA BOOK. St. Paul: 1975.

Mississipi. State University. College of Business and Industry. Division of Research. MISSISSIPPI STATISTICAL ABSTRACT. State College: 1976.

Missouri. Energy Agency. MISSOURI ENERGY PROFILES. Jefferson City: 1976.

_____. 1976 ENERGY RESEARCH AND DEVELOPMENT INVENTORY FOR THE STATE OF MISSOURI. Jefferson City: 1976.

Missouri. University of. Extension Division. DATA FOR MISSOURI COUNTIES. Columbia, Mo.: 1970-- . Updated periodically. Loose-leaf.

Mitchell, B.R. EUROPEAN HISTORICAL STATISTICS, 1750-1970. New York: Columbia University Press, 1975.

> Comparable to the HISTORICAL STATISTICS OF THE UNITED STATES FROM COLONIAL TIMES TO 1970, this publication provides time series data on the output of coal, crude petroleum, natural gas, and imports and exports.

Montana. Energy Advisory Council. MONTANA HISTORICAL ENERGY STATISTICS. Helena: 1976.

Montana. State Division of Research and Information Systems. MONTANA DATA BOOK. Helena: 1970-- . Updated periodically. Loose-leaf.

Moody's Investor's Services, Inc. MOODY'S INDUSTRIAL MANUAL. New York: 1914-- . Annual, with updates.

Investment informatory service containing financial and related data on manufacturing companies in the field of energy.

Mutch, James J. TRANSPORTATION ENERGY USE IN THE UNITED STATES: A STATISTICAL HISTORY, 1955-1971. Report no. R-1391. Santa Monica: RAND, 1973.

Myers, John. INDUSTRIAL ENERGY DEMAND, 1976-2000. Washington, D.C.: U.S. General Accounting Office, 1969.

_____. SAVING ENERGY IN MANUFACTURING: THE POST-EMBARGO RECORD. Cambridge, Mass.: Ballinger, 1978.

National Science Foundation. Oak Ridge National Laboratory. Environmental Program. THE ENERGY COST OF GOODS AND SERVICES. By Robert A. Herendeen. Oak Ridge, Tenn.: 1973.

Calculates total energy costs of many consumer goods and services. The energy types included in the report are coal, crude, oil, and gas extraction, refined petroleum, electricity, and gas sales. Statistical data include the following: U.S. primary energy use by consuming sectors, energy allocation and pricing, detailed energy inverse pricing, breakdown of total energy requirements for selected products, comparison of dollar and electrical energy input-output coefficients for selected sectors, energy use versus GNP, and energy impact of the automobile.

National Science Teachers Association. ENERGY ENVIRONMENT SOURCE BOOK. 2 vols. Washington, D.C.: 1975.

Intended as an educational aid for science teachers. The first volume is entitled ENERGY SOCIETY AND THE ENVIRONMENT and the second, ENERGY, ITS EXTRACTION CONVERSION AND USE. The volumes include the following data: categories of energy consumption, and uses, sources of energy and pollution, energy-to-GNP ratio, household energy use by income group, average output versus energy input to food systems, and distribution of total environmental expenditures.

Nebraska. Department of Economic Development. Division of Research. NEBRASKA STATISTICAL HANDBOOK. Lincoln: 1977.

Nevada. Department of Economic Development. NEVADA COMMUNITY PROFILES. Carson City: 1970.

New Hampshire. Department of Resources and Economic Development. NEW HAMPSHIRE ECONOMIC INDICATORS. 1974.

New Jersey. Office of Business Economics. COUNTY DATA SUMMARY. Trenton, N.J.: 1975.

New Mexico. University of. Bureau of Business Research. NEW MEXICO STATISTICAL ABSTRACT, VOL. 3. Albuquerque: 1975.

New Orleans. University. Division of Business and Economic Research. STATISTICAL ABSTRACT OF OKLAHOMA. Norman: 1976.

New York. Division of Budget. Office of Statistical Coordination. NEW YORK STATE STATISTICAL YEARBOOK. Albany: 1969-- .

North Carolina. Department of Administration. Office of the State Budget, in cooperation with Association for Coordinating Interagency Statistics. NORTH CAROLINA STATE GOVERNMENT STATISTICAL ABSTRACT. Raleigh: 1976.

North Carolina. Department of Veterans and Military Affairs. Energy Division. Research Division. ENERGY CONSUMPTION IN NORTH CAROLINA. Raleigh: 1974-- . Annual.

North Dakota. Business and Industrial Development Department. NORTH DAKOTA GROWTH INDICATORS. Bismarck: 1976.

Ohio. Department of Economic and Community Development. Office of Population Statistics. STATISTICAL ABSTRACT OF OHIO. 3d ed. Columbus: 1977.

OIL WORLD STATISTICS. London: Institute of Petroleum, Info-Services Department, 1975.

Oklahoma. University of. Center for Economic and Management Research. STATISTICAL ABSTRACT OF OKLAHOMA. Norman: 1976.

Oregon. University of. Bureau of Business Research. OREGON ECONOMIC STATISTICS. Rev. ed. Eugene: 1977.

Organization for Economic Cooperation and Development. BASIC ENERGY STATISTICS AND ENERGY BALANCES OF DEVELOPING COUNTRIES, 1967-1977. Paris: 1979.

_____. ENERGY BALANCES OF OECD COUNTRIES. Paris: 1976-- . Annual.

A comprehensive presentation of basic statistics in original units, such as tons of coal and kilowatt hours of electricity, a statistic required for analysis of energy policy problems. This volume represents the first major attempt by the OECD to meet such requirements, and provides standardized energy balance sheets expressed in a common unit of tons of oil equivalent for all OECD countries for years under study.

_____. ENERGY PROSPECTS TO 1985. 2 vols. Paris: 1975.

_____. HISTORICAL ENERGY STATISTICS. Paris: 1977.

Covers the years 1960–75 and provides composite data on energy for OECD countries.

_____. REFINERY CAPACITY. Paris: 1977.

Data on capacity and utilization trends of refineries in OECD countries.

_____. STATISTICS OF FOREIGN TRADE. Paris: 1955-- . Monthly.

Data on total OECD trade by country of origin, area of origin, destination, and major commodity. Includes energy–related commodities. The publishers also issue a quarterly series by country and market, called TRADE BY COMMODITIES.

_____. WORLD ENERGY OUTLOOK. Paris: 1977.

Gives projected energy supply and demand to 1985 for OECD countries to be used in examining policy options to lower the import needs of industrial countries.

Parra, Ramos, and Parra Ltd. of Venezuela. Consultants in Petroleum Economics. WORLD SUPPLIES OF PRIMARY ENERGY, 1976-1980. Wokingham, Engl.: Energy Economics Information Service, 1976.

This new study examines the year–by–year development of primary energy supplies to 1980. Detailed estimates by fuel and by major country and region are based on work in progress or planned and lead times in the energy industries.

Pennsylvania. Department of Commerce. Bureau of Statistics. Research and Planning. PENNSYLVANIA STATISTICAL ABSTRACT. Harrisburg: 1959-- . Annual.

Petro Canada. INTERNATIONAL ENERGY PROSPECTS TO 2000. Ottawa: 1978.

Potter, Neal, and Christy, Francis T. TRENDS IN NATURAL RESOURCE

COMMODITIES. Baltimore: Published for the Resources for Future by Johns
Hopkins Press, 1962.

> Historical study of the energy resources industry. Includes a
> good deal of statistical data, as follows: prices, output, con-
> sumption, foreign trade, and employment in the energy field in
> the United States from 1870, with projection to 1975.

Predicasts, Inc. PREDICASTS BASEBOOK. Cleveland: 1973-- . Annual.

> Also arranged by an eight-digit standard industry code, the statis-
> tical abstract gives historical annual time series data on energy
> and related industries. Details include P & E expenditures, pay-
> rolls, employment, weekly work hours, and annual rate of growth.

_____. WORLD ENERGY SUPPLY AND DEMAND. World Studies Series,
no. 93. Cleveland: 1974.

> This report investigates the world supply and demand for energy.
> Forty countries and/or geographical areas are covered. For each
> country, historical data from 1961 and projections for 1960 and
> 1985 are provided for the four primary sources of energy, that
> is, coal, natural gas, oil products and nonhydrocarbons, and the
> four major endusers, that is, industry, motor vehicles, households,
> and electric power utilities.

_____. WORLD ENERGY SUPPLY AND DEMAND. World Studies Series,
no. 140. Cleveland: 1977.

> This report analyzes world primary energy requirements and sup-
> plies, with emphasis on the long-term outlook to 1990 as well as
> for 1980 and 1985. Data on energy consumption and production
> since 1960 for over thirty-five countries and regions are pre-
> sented. Primary energy forms include coal, crude petroleum,
> natural gas, hydroelectricity, and nuclear and advanced power
> sources. Demand sectors are motor vehicles, households, indus-
> try, and electricity generation.

Puerto Rico. Planning Board. Bureau of Statistics. STATISTICAL YEARBOOK.
Santurce: 1976.

Ray, George, and Robinson, Colin. THE EUROPEAN ENERGY MARKET TO
1980. London: Stanliland Hall Associates, 1975. U.S. distributor, Bookstax
of Britain, Ltd., New York.

> Analyzes the consequences of the Arab oil price increase for
> energy markets. It forecasts the price of crude oil for each
> year into the 1980s, and the demand and import requirements
> for energy for each European country to 1980. There are twenty-
> two tables covering the following countries: Belgium, Denmark,

France, West Germany, Ireland, Italy, Netherlands, United Kingdom, Norway, and Sweden.

Resource Programs, Inc. PROFILE: A CONTINUING STUDY OF OIL AND GAS INDUSTRY. New York: 1970-- . Irregular.

> Statistical descriptions of public oil and gas exploration programs taken from reports filed with the U.S. Securities and Exchange Commission. Data include cost of the project and sponsors.

_____. RPI COMPENDIUM OF OIL AND GAS PROGRAMS. New York: 1970-- . Annual.

> Companion volume to the PROFILE (see above), containing similar information on publicly offered programs.

Rhode Island. Department of Economic Development. RHODE ISLAND BASIC ECONOMIC STATISTICS. Providence: 1977.

Robert Morey Associates. ENERGY INFORMATION. Dana Point, Calif.: 1965-- . Monthly.

> Data on research and development activities in the energy industry, contracts and grants, new products and technologies, company information, and power plant construction.

Rockefeller Foundation. INTERNATIONAL ENERGY SUPPLY: A PERSEPCTIVE FROM THE INDUSTRIAL WORLD. New York: 1978.

Ross, H. John, ed. THE ENERGY ACTIVISTS DIRECTORY. Miami: Office Research Institute, 1975.

> This 110-page directory is basically a who's who on energy conservation, conversion, alternatives, power resources, and environmental protection.

Schurr, Sam H., and Netschert, Bruce C. ENERGY IN THE AMERICAN ECONOMY, 1850-1975. Baltimore: Published for the Resources for Future by Johns Hopkins Press, 1960.

> An economic study of the energy industry, its history and prospects. Like any other economic history study, this contains vast amounts of statistical data compiled from numerous sources.

SECURITY PRICE INDEX RECORD. New York: Standard and Poor's Corp., 1962-- . Annual.

> Contains historical time series data on the daily and weekly stock price indicators for industry groups, including petroleum and coal as well as utility stocks.

Shell Oil Company. THE NATIONAL ENERGY OUTLOOK, 1980-1990. Houston, Tex.: 1976.

Societe de Documentation et d'Analyses Financieres. INFORMATIONS IN-TERNATIONALES. Paris: 1970-- .

> Like JANE'S MAJOR COMPANIES OF EUROPE (see above, under Charter House Group), this continually updated publication provides information on power, oil, and mineral companies in Western Europe, United States, Brazil, Uruguay, and the Union of South Africa.

South Carolina. Budget and Control Board. Division of Research and Statistical Services. ECONOMIC REPORT FOR SOUTH CAROLINA: Columbia: 1976.

_____. SOUTH CAROLINA STATISTICAL ABSTRACT. Columbia: 1977.

South Dakota. State Planning Bureau. SOUTH DAKOTA FACTS. Pierre: 1976.

South Dakota. University. Business Research Bureau. SOUTH DAKOTA ECONOMIC AND BUSINESS ABSTRACT. Vermillion: 1972.

Standard and Poor's Corp. STOCK REPORTS. New York: 1973 - . Quarterly.

> Financial and related statistical data on all energy companies listed on national and regional stock exchanges. Also available loose-leaf.

Statistical Office of the European Communities. EUROSTAT-ENERGY STATIS-TICS. Brussels, Belgium: 1976-- . Annual.

> Covers the following countries: France, West Germany, Italy, Netherlands, Belgium, Luxembourg, United Kingdom, Ireland, and Denmark. Contains data on the following: energy overall balance, solid fuels (hard coal, patent fuel coke, lignite) petroleum products and gas, natural gas, derived gases, electrical energy, energy economics, product consumption, import and export investment, prices, and refineries.

Tennessee. University of. Center for Business and Economic Research. TENNESSEE STATISTICAL ABSTRACT. 3d ed. Knoxville: 1974. Triennial.

_____. TENNESSEE POCKET DATA BOOK. 3d ed. Knoxville: 1975.

Texas Eastern Transmission Corp. COMPETITION AND GROWTH IN AMERI-
CAN ENERGY MARKETS, 1974-1985. Houston: 1968.

> This study analyzes the structural and competitive relationships
> between fuels and their markets. A significant amount of statis-
> tical data is used to support the study, including the actual and
> projected requirements, sources of supply of total end-use energy
> and composition of end-use energy requirments and sources of
> supply.

Texas. Office of Information Services. Management Science Division.
ENERGY SUPPLY AND DEMAND IN TEXAS. Austin: 1974.

Trends Publishing, Inc. ENERGY TODAY. Washington, D.C.: 1973-- .
Semimonthly.

> Information on the economic aspects of the energy industry, new
> technologies, energy use patterns, supply and demand, new
> sources, and research activity conducted by private and govern-
> ment agencies.

United Nations. Statistical Office. ANNUAL BULLETIN OF GENERAL
ENERGY STATISTICS FOR EUROPE. New York: 1968-- .

> Deals with the production of primary and derivative energy for
> solid fuels, liquid fuels, gaseous fuels, electric energy, steam
> and hot water, imports, exports, bunkering, gross consumption,
> energy converted, consumption for non-energy-users, net con-
> sumption, losses in transport and distribution, and breakdown of
> consumption by energy-producing industries.

_____. COMMODITY TRADE STATISTICS. New York: 1962-- . Quarterly.

> Quarterly statistics on world commodity trade by region, by
> country of provenance, and by destination. Includes energy-
> related commodities.

_____. ENERGY ATLAS FOR ASIA AND THE FAR EAST. New York: 1970.

_____. MONTHLY BULLETIN OF STATISTICS. New York: 1947-- .

> Supplements the United Nations STATISTICAL YEARBOOK (see
> below), updating its section on energy.

_____. STATISTICAL YEARBOOK. New York: 1948-- .

> Contains data on output and employment in electricity, gas and
> water supply, energy production and consumption, installed capa-
> city, electric energy, and manufactured gas production.

_____. WORLD ENERGY SUPPLIES, 1950-1974. Series J, no. 19. New York: 1976.

This is a special historical study of an annual series under the same title. Contains the following international statistical information for the period concerned: production, consumption, and trade of commercial energy, solid-fuels, crude petroleum and products, LP gases, electrical energy, and nuclear fuels. Listed by country or area.

_____. YEARBOOK OF INDUSTRIAL STATISTICS. New York: 1967-- .

Formerly called the Growth of World Industry, this report gives for each country data on its coal, petroleum, gas, metal ore, mining, quarrying, electric, gas, and steam utilities. Details on each industry include the number of establishments, employment, wage and salaries of operatives, gross output in factor values, value added in factor values, gross fixed capital formation, value of the stocks at the end of the period, and index numbers of industrial production.

_____. YEARBOOK OF INTERNATIONAL TRADE. New York: 1950-- .

Arranged by country and commodity, this report contains data on world export and import trade in energy-related commodities.

U.S. Board of Governors of the Federal Reserve System. INDUSTRIAL PRODUCTION. Washington, D.C.: Government Printing Office, 1943-- . Biennial.

Provides indexes of industrial production, output, and market structure of the American economy. It includes coal, coke and products, crude oil and natural gas, gas and electric utilities, and petroleum refining and products. The 1971 volume is supplemented by data in the monthly FEDERAL RESERVE BULLETIN.

U.S. Bureau of Economic Analysis. BUSINESS STATISTICS. Washington, D.C.: Government Printing Office, 1947-- . Biennial.

Puts together historical data from 1947 to 1974 for the 2,500 series that appear in the statistical section of the SURVEY OF CURRENT BUSINESS (see below). Includes energy-related industries.

_____. REGIONAL EMPLOYMENT BY INDUSTRY, 1940-1970. Washington, D.C.: Government Printing Office, 1975.

Issued as a special supplement to the SURVEY OF CURRENT BUSINESS (see below), this report provides decennial series employment data for the United States, by state, region, and country.

Includes data on county employment in energy-related occupations in mining, manufacturing, utilities, wholesale and retail trade, and services.

_____. SURVEY OF CURRENT BUSINESS. Washington, D.C.: Government Printing Office, 1921-- . Monthly.

Contains energy-related economic statistics, updated by the weekly supplement, also called BUSINESS STATISTICS.

U.S. Bureau of Labor Statistics. EMPLOYMENT AND EARNINGS. Washington, D.C.: Government Printing Office, 1954-- . Monthly, with annual cumulation.

This bulletin gives labor statistics by standard industrial product classification. The following industries are included: bituminous coal and lignite mining, oil and gas extraction, oil and gas field services, petroleum refining, pipeline transportation, electric and gas utilities, and gasoline service stations. There is a companion annual publication called EMPLOYMENT AND EARNINGS, STATES AND AREAS, which gives similar data by geographical breakdowns.

U.S. Bureau of Labor Statistics. Office of Prices and Living Conditions. CONSUMER PRICE INDEX DETAILED REPORT. Washington, D.C.: Government Printing Office, 1941-- . Monthly.

Includes price indexes for fuel and utilities, fuel oil and coal, gas and electricity, gasoline and motor oil.

_____. RETAIL GASOLINE PRICES. Washington, D.C.: Government Printing Office, 1952-- .

Periodical release containing price indexes and average retail prices of regular and premium gasoline for the United States and twenty-three selected cities.

_____. WHOLESALE PRICES AND PRICE INDEXES. Washington, D.C.: Government Printing Office, 1956-- . Monthly.

Contains price indexes and average prices for various energy products including anthracite, natural gas, propane, electricity, gasoline, crude petroleum, and residual fuel oil.

U.S. Bureau of Mines. COMMODITY DATA SUMMARIES. Washington, D.C.: Government Printing Office, 1976-- . Annual.

Report on anthracite, bituminous coal and lignite coal, natural gas, natural gas liquids, petroleum, and uranium. All these commodities are examined from the point of view of production

per man, stocks, employment, tariff, depletion allowances, and
government programs.

_____. MINERAL FACTS AND PROBLEMS. Bulletin no. 650. Washington,
D.C.: Government Printing Office, 1975.

Review of energy resources in the areas of anthracite, lignite
and bituminous coal, carbon, helium, hydrogen, natural gas,
peat, petroleum, shale oil, thorium, and uranium. It provides
data on the production and demand for those minerals, forecasts
to the year 2000, estimates of the resources, consumption by
major sources, industry patterns, technology, reserves, prices,
costs, employment, and productivity.

_____. MINERALS AND MATERIALS: A MONTHLY SURVEY. Washington,
D.C.: Government Printing Office, 1976-- . Monthly.

Energy commodities analyzed in this monthly report include the
following: crude petroleum, petroleum, natural gas liquids,
bituminous coal, lignite and anthracite, and fossil fuels. Statistics collected cover production consumption, exports, imports,
inventories, and price.

_____. MINERALS YEARBOOK. 2 vols. Washington, D.C.: Government
Printing Office, 1932-- .

A comprehensive source of information on energy-related minerals
both within and outside the United States, containing, among
other things, the following data: production and value of natural
gas liquids at plants and liquefied petroleum gases by state, value
of all petroleum minerals as a percent of total value of all minerals produced in leading oil states, natural gas processing plant
operations in the United States, and gross consumption (in BTUs)
of energy resources by major source and coal mining sector.

_____. STATUS OF MINERAL INDUSTRIES. Washington, D.C.: Government
Printing Office, 1976-- . Annual.

Covers mining, minerals, metals, and mineral reclamation.
Charts describe requirements for each U.S. city and role of these
minerals in the industry. Also includes data on U.S. production
in relation to the rest of the world; imports by major foreign
sources; consumption by source, and financial data. The cover-
age includes petroleum, coal, natural gas, and uranium.

U.S. Bureau of Mines. Minerals and Materials Supply Demand Analysis.
MINERALS IN THE U.S. ECONOMY. TEN-YEAR SUPPLY DEMAND PROFILE
FOR MINERALS AND FUEL COMMODITIES. Washington, D.C.: Government
Printing Office, 1975.

Staff report covering bituminous coal, natural gas, petroleum, uranium, and cobalt. It gives data on world and U.S. production of all the above fuels, components of U.S. supply in terms of domestic mines and imports and distribution of U.S. demand pattern by type of use.

U.S. Bureau of the Census. CENSUS OF BUSINESS. Washington, D.C.: Government Printing Office, 1972.

Comprehensive statistics on business establishments in energy-related, wholesale trade, and selected service industries. The data include the nature of business, location, legal form of organization, size, payroll and employment, sales or receipts, and sales by merchandise lines. Fuel oil dealers, petroleum gas dealers, and gas service stations fall within the purview of the business census. The data are available for the United States, all the states, and standard metropolitan statistical areas. It is updated by monthly and weekly trade reports from the Bureau of Census.

_____. CENSUS OF HOUSING. Washington, D.C.: Government Printing Office, 1940-- . Decennial.

Detailed data on housing characteristics in the United States including heating and air conditioning facilities.

_____. CENSUS OF MANUFACTURERS. Washington, D.C.: Government Printing Office, 1932-- . Quinquennial.

Energy-related manufacturing industries are included within the scope of this census, containing the following information: number of establishments, quantity and value of products shipped, quantity and cost of materials consumed, cost of fuels and electric energy, capital expenditures, values of fixed assets, rental payments for buildings, and value added. The quinquennial census is updated by the ANNUAL SURVEY OF MANUFACTURERS and the CURRENT INDUSTRIAL REPORTS.

_____. CENSUS OF MINERAL INDUSTRIES. Washington, D.C.: Government Printing Office, 1972.

Contains industry statistics such as oil and natural gas offshore operations by geographic division, financial statistics of companies engaged in the exploration and development of crude petroleum and natural gas by area, companies engaged in natural gas liquids, the crude oil and natural gas well drilling service industry, petroleum refining industry and employment, and payroll statistics of these companies.

_____. CONSTRUCTION EXPENDITURES OF STATE AND LOCAL GOVERN-

MENTS. Government Census 75, no. 2. Washington, D.C.: Government Printing Office, 1970-- . Quarterly.

Includes spending volumes on electric power systems.

_____. COUNTRY BUSINESS PATTERNS. Washington, D.C.: Government Printing Office, 1943-- . Annual.

Data on total number of establishments, employment size and payroll of businesses, including those related to energy, arranged by county and state, with a section on the entire nation.

_____. CURRENT CONSTRUCTION REPORTS. CHARACTERISTICS OF NEW ONE-FAMILY HOMES. C-25 series. Washington, D.C.: Government Printing Office, 1959-- . Monthly.

Describes house heating fuel and air conditioning systems.

_____. HISTORICAL STATISTICS OF THE UNITED STATES. COLONIAL TIMES TO 1970. Washington, D.C.: Government Printing Office, 1976.

Contains more than 12,500 time series on various topics, including energy.

_____. STATISTICAL ABSTRACT OF THE UNITED STATES. Washington, D.C.: Government Printing Office, 1878-- . Annual.

One section of this indispensable publication is devoted to energy and power resources in the United States.

U.S. Central Intelligence Agency. HANDBOOK OF ECONOMIC STATISTICS. Washington, D.C.: Document Expediting Project of the Library of Congress (DOCEX), 1976-- . Annual.

Covers all Communist and selected non-Communist countries. Contains data on production of primary energy, by type, proved reserves of crude oil, natural gas, coal, crude oil refining capacity, estimated oil trade, installed electric power generating capacity, and nuclear power generating capacity.

_____. THE INTERNATIONAL ENERGY SITUATION: OUTLOOK TO 1985. Washington, D.C.: Government Printing Office, 1977.

U.S. Central Intelligence Agency. Office of Economic Research. INTERNATIONAL OIL DEVELOPMENTS: STATISTICAL SURVEY. Washington, D.C.: Document Expediting Project of the Library of Congress (DOCEX), 1976-- . Biweekly.

Contains data on world crude production, world natural gas

liquid production, OPAEC and OPEC countries crude oil production and capacity-estimated proved and provable petroleum reserves, crude oil imports of selected developed countries by source; trends in oil trade of selected developed countries; and price structure.

U.S. Congress. House. Banking and Currency Committee. ENERGY SECURITY AND THE DOMESTIC ECONOMY: IMPACT ON PRICES, EMPLOYMENT AND CONSUMPTION. 93d Cong., 2d sess. Washington, D.C.: 1974.

Report of the ad hoc committee on the domestic and international monetary effect of energy and other natural resource pricing, containing an analysis of the impact on the national economy through 1985 of higher energy prices.

U.S. Congress. House. Committee on Interstate and Foreign Commerce. BASIC ENERGY DATA AND GLOSSARY OF TERMS. Washington, D.C.: Government Printing Office, 1975.

U.S. Congress. House. International Relations Committee. Subcommittee on International Resources, Food and Energy. U.S. INTERNATIONAL ENERGY POLICY. Washington, D.C.: Government Printing Office, 1975.

Statistical data on oil and other commodity prices, price indexes, and trade, collected mostly from international agencies.

U.S. Congress. Senate. Committee on Finance. ENERGY STATISTICS. 94th Cong., 1st sess. Washington, D.C.: Government Printing Office, 1975.

This is an informational report containing alternative estimates of U.S. undiscovered recoverable petroleum resources, natural gas reserves and estimates, demonstrated U.S. coal reserve bases, world recoverable energy reserves, and world proven reserves.

U.S. Congress. Senate. Judiciary Committee. Subcommittee on Antitrust and Monopoly. INTERFUEL COMPETITION. 94th Cong., 1st sess. Washington, D.C.: Government Printing Office, 1976.

Contains statistical data on natural gas or petroleum, producers and refiners who own interest in other energy industries.

U.S. Council of Environmental Quality. ENERGY ALTERNATIVES: A COMPARATIVE ANALYSIS. Washington, D.C.: Government Printing Office, 1975.

Report on data and methods for computing economic and environmental costs of alternative energy technologies utilizing fossil, nuclear, solid waste, and various nonexhaustible resources such as hydroelectric and solar power.

U.S. Department of Commerce. CONSTRUCTION VOLUME AND COSTS

1915-1956. A STATISTICAL SUPPLEMENT TO CONSTRUCTION REVIEW. Washington, D.C.: Government Printing Office, 1958.

A joint publication of the departments of commerce and labor, this report presents monthly and annual data on construction put in place by type, including petroleum pipelines; and electric light, power, and gas utilities.

U.S. Department of Energy. ENERGY AVAILABILITIES FOR STATE AND LOCAL DEVELOPMENT. Springfield, Va.: National Technical Information Service, 1976-- . Annual.

Report estimating the demand, production, and imports of electricity and seven fossil fuel products for 9 census divisions, 50 states and 173 BEA regions.

_____. PROJECTIONS OF ENERGY SUPPLY AND DEMAND AND THEIR IMPORTS. 2 vols. Washington, D.C.: 1978.

_____. RESIDENTIAL ENERGY USE TO THE YEAR 2000: A REGIONAL ANALYSIS. By Eric Hirst and James B. Kurish. Springfield, Va.: National Technical Information Service, 1977. 47 p.

Report on residential energy use, costs and savings potential under four federally authorized conservation programs, by region, 1977-2000.

U.S. Department of the Interior. ANNUAL REPORT OF THE SECRETARY OF THE INTERIOR UNDER THE MINING AND MINERALS POLICY ACT OF 1970. [public law 91-631]. Washington, D.C.: Government Printing Office, 1971-- .

The report, together with its statistical appendix, contains detailed profiles of the mineral industries of the United States. The energy-related minerals covered in the reports are coal, cobalt, natural gas, petroleum, shale oil, and uranium. For each mineral analyzed the exploration and resource positions in the United States are given.

_____. ENERGY PERSPECTIVES. Washington, D.C.: Government Printing Office, 1975.

A presentation of major energy and energy-related data by type, historical and current, as below: world recoverable energy reserves by source, energy consumption by source, U.S. recoverable energy resources, energy end uses, consumption patterns, per capita consumption, and projections to 1990.

_____. ENERGY PERSPECTIVES 2. Washington, D.C.: Government Printing Office, 1976.

Sequel to the volume published under similar title in 1975 (see above). Includes new and revised energy data. The aim is to portray the basic parameters affecting the U.S. energy picture as well as to present additional data integral to statistical, econometric, and other analyses.

_____. UNITED STATES ENERGY THROUGH THE YEAR 2000. Prepared by Walter G. Dupree and James A. West. Rev. ed. Washington, D.C.: Government Printing Office, 1976.

Forecasts energy consumption within each major consuming sector and source of energy supplies through 2000. Individual forecasts are available for total energy; electric power; synthetic gas; household, commercial, transportation, industrial usage; petroleum; gaseous fuel; and coal.

U.S. Department of the Interior. Bureau of Land Management. ENERGY ALTERNATIVES: A COMPARATIVE ANALYSIS. Washington, D.C.: 1973-- . Annual.

Contains data on major energy resource systems in the United States, efficiencies, environmental impact, and costs.

_____. OUTER CONTINENTAL SHELF STATISTICAL SUMMARY. Washington, D.C.: Government Printing Office. Covers years 1973-75. Frequency varies.

Contains data on bids for energy exploration on government land and property by various companies. It gives details like name of company, amount of exploration for each company, bid per acre, total bid, standing of the bid, and total area to be bid.

U.S. Domestic and International Business Administration. U.S. INDUSTRIAL OUTLOOK, WITH PROJECTIONS TO 1990. Washington, D.C.: Government Printing Office, 1960-- . Annual.

Current and projected data on all energy-related industries.

U.S. Energy Information Administration. ANNUAL REPORT TO THE CONGRESS. Washington, D.C.: Government Printing Office, 1978-- .

Presents data on energy production, consumption, and prices, 1947-77, and projections for selected years 1978-90.

_____. END USE ENERGY CONSUMPTION DATA BASE. Springfield, Va.: National Technical Information Service, 1978. 206 p.

Estimates energy consumption for agriculture, mining, manufacturing, construction, transportation, and electric utilities.

_____. ENERGY AND COST ANALYSIS OF RESIDENTIAL HEATING SYSTEMS.

Springfield, Va.: National Technical Information Service, 1978. 59 p.

    Estimates cost effectiveness and energy conservation potential of design improvements in residential space heating systems.

_____. FEDERAL ENERGY DATA SYSTEM (FEDS) STATISTICAL SUMMARY. Washington, D.C.: Government Printing Office, 1978-- . Annual.

    Presents detailed energy consumption data for 1960-75, by consuming sector, energy source, region and state.

U.S. Energy Research and Development Administration. FUEL CHOICES IN THE HOUSEHOLD SECTOR. By William Lin et al. Springfield, Va.: National Technical Information Service, 1976.

    Report on residential fuel choice as a function of fuel and household equipment prices, income, and selected demographic and climatic variables, in the following areas: home heating, cooking, air conditioning, and food freezing.

_____. NATIONAL PLAN FOR ENERGY RESEARCH DEVELOPMENT AND DEMONSTRATION. CREATING ENERGY CHOICES FOR THE FUTURE. Vol. 1, PLAN; vol. 2, PROGRAM IMPLEMENTATION. Washington, D.C.: Government Printing Office, 1976.

    Recommends energy research and development goals and objectives, in terms of major energy technology, options, and potential energy contribution. Discusses federal and private-sector roles and year-to-year objectives for energy independence and estimates.

_____. SAVINGS IN ENERGY CONSUMPTION BY RESIDENTIAL HEAT PUMPS: THE EFFECTS OF LOWER INDOOR TEMPERATURES AND OF NIGHT SETBACK. By R.D. Ellison. Springfield, Va.: National Technical Information Service, 1977.

    Estimates the potential household energy savings from operating residential heat pumps at various thermostat settings in six cities.

U.S. Environmental Protection Agency. COMPARISON OF FOSSIL AND WOOD FUELS. Prepared by E.H. Hall et al. Publication no. PB 251 622. Springfield, Va.: National Technical Information Service, 1976.

    Contains data on the costs and benefits of using wood rather than oil or coal for fueling a planned 50-megawatt electric power plant in central Vermont.

_____. ENERGY CONSUMPTION FUEL UTILIZATION AND CONSERVATION IN INDUSTRY. By John T. Reding and Burchard P. Shepherd. Springfield, Va.: National Technical Information Service, 1975.

    Report on fuel utilization, heat rejection, and potential effects of energy conservation actions in six industry groups: (1) chemicals; (2) primary metals; (3) petroleum; (4) paper; (5) stone, clay, glass, and concrete; and (6) food.

U.S. Executive Office of the President. ENERGY ALTERNATIVES: A COMPARATIVE STUDY. Washington, D.C.: Government Printing Office, 1975.

Contains descriptive data for all major energy alternatives that the United States might develop over the next twenty five years, including coal, oil shale, petroleum, natural gas, tar sands, hydroelectric and nuclear energy sources, and emerging energy technologies such as geothermal energy and organic waste conversion.

U.S. Federal Energy Administration. ANNUAL REPORT. Washington, D.C.: 1972-- .

Includes an overview of the nation's energy supply outlook 1975-85 in terms of the following: projected production and consumption levels to 1985 for coal, electricity (including nuclear-generated), natural gas, and petroleum products; and projected constraints in finance, labor, and equipment.

_____ . COMPARISON OF ENERGY CONSUMPTION BETWEEN WEST GERMANY AND THE U.S. By Richard L. Goon and Ronald K. White, Stanford Research Institute. Washington, D.C.: 1976.

Report comparing per capita and end-use energy consumption in the United States and West Germany in 1972, by major economic sector.

_____ . DIRECTORY OF FEDERAL ENERGY DATA SERVICES: COMPUTER PRODUCTS AND RECURRING PUBLICATIONS. Springfield, Va.: National Technical Information Service, 1976. Annual.

Identifies two major types of federally sponsored energy-related information: (1) energy information on magnetic tape, (2) recurring publications that contain energy-related numerical data. It is indexed by agency and subject.

_____ . ECONOMIC IMPACT OF ENERGY SHORTAGES ON COMMERCIAL AIR TRANSPORTATION AND AVIATION MANUFACTURE. Springfield, Va.: National Technical Information Service, 1975.

Analysis of commercial air carriers' and aviation manufacturers' status and future prospects to 1980-90 in an energy shortage situation, in terms of direct and indirect effects of energy shortages and increased fuel costs. Includes detailed profiles of industry, carrier, and manufacturing sectors.

_____. ENERGY AND U.S. AGRICULTURE. Washington, D.C.: Government Printing Office, 1976.

Contains data on the use and conservation of energy in agricultural production.

_____. ENERGY CONSERVATION POTENTIAL IN THE CEMENT INDUSTRY. Conservation paper, no. 26. Washington, D.C.

Report by the Portland Cement Association on energy use in Portland cement production, and energy intensiveness of existing and new generation production equipment and processes in the United States, Canada, and Europe. Includes historical data and projections to 2000.

_____. ENERGY INFORMATION. Washington, D.C.: Government Printing Office, 1976-- . Quarterly.

This is a report on the energy situation submitted to the Congress as required by Public Law 93-319. Provides data on natural gas production, average daily productive capacity, underground storage, consumption, marketed production, average runs to stills, fossil fuels, imports and exports, and geophysical exploration activity.

_____. FEDERAL ENERGY MANAGEMENT PROGRAM. Washington, D.C.: Government Printing Office, 1976-- . Quarterly, with annual cumulation.

This is a report on the efforts to conserve federal use of energy. It provides data on the energy savings achieved by each agency, consumption and saving by source, and energy conservation in vehicles and equipment operations.

_____. IMPACT OF ENERGY PRICE INCREASES ON LOW INCOME FAMILIES. Springfield, Va.: National Technical Information Service, 1976.

Report by Mathematica, Inc., calculating average household fuel expenditures, and increases in fuel expenditures as a percent of disposable income, by income class and region. Covers expenditures for gasoline, for home use of piped natural gas, bottled gas or liquid petroleum gas, fuel oil, coal, and electricity.

_____. NATIONAL ENERGY OUTLOOK. Washington, D.C.: Government Printing Office, 1976-- . Annual.

Contains energy forecasts through 1990 based upon national energy supply and demand models developed by the FGA during the previous two years. Forecasts are also made for prices, consumption, supply, emerging technologies, import situation, capital requirements, environmental aspects, oil, electric utilities, natural gas, nuclear fuel cycle, coal, and synthetic fuels.

_____. PROJECT INDEPENDENCE. 30 vols. Washington, D.C.: Govern-
Printing Office, 1975.

> A definitive study of the energy sector from the point of view of
> resources, needs, and potentials. It contains a wealth of energy-
> related statistical data on all aspects of energy.

_____. REPORT TO CONGRESS: ENERGY CONSERVATION STUDY.
Washington, D.C.: Government Printing Office, 1974.

> Contains statistical analysis of the energy conservation implica-
> tions of U.S. exports of major fuel commodities and energy-
> intensive products. Also deals with energy savings potential
> in industry, 1980–85, by means of increased recycling and other
> programs.

U.S. Federal Energy Administration. Office of Data and Analysis. ENERGY
INFORMATION IN THE FEDERAL GOVERNMENT. Washington, D.C.: Government Printing Office, 1976.

> Directory of energy sources identified by the Interagency Task
> Force on Energy Information. It contains a Federal Energy
> Information Locator System, which is a directory to most of the
> federal government sources of energy information. Provides a
> description of various programs, major energy data activities, and
> an index of programs by source.

U.S. Federal Power Commission. STATE PROJECTIONS OF INDUSTRIAL FUEL
NEEDS. By Marquis R. Seidel. Washington, D.C.: 1976.

> Estimates of the impact of increased fuel prices on the manufactur-
> ing industry's energy consumption and earnings in 1980 and 1990,
> by state, region, and industry group.

U.S. Federal Trade Commission. CONCENTRATION LEVELS AND TRENDS IN
THE ENERGY SECTOR OF THE UNITED STATES ECONOMY. Washington,
D.C.: Government Printing Office, 1974.

> Staff economic report on the competition levels and trends in the
> crude oil, natural gas, coal, and uranium industries and the
> diversification patterns. It provides information on the largest
> energy holders, largest energy producers, regional production and
> distribution, and prices and market shares.

_____. QUARTERLY FINANCIAL REPORT FOR MANUFACTURING, MINING
AND TRADE CORPORATIONS. Washington, D.C.: Government Printing Office, 1947-- .

> Contains business and financial ratios, composite income state-
> ment, balance sheets, and rates of return data, classified by
> industry and asset size. Includes petroleum and coal companies.

U.S. Geological Survey. Conservation Division. OUTER CONTINENTAL SHELF STATISTICS. Washington, D.C.: Government Printing Office, 1976.

> Gives oil and gas leasing, drilling production, and income statistics. Includes state by state statistics on exploration, number of wells drilled and completed, number of producing wells, and employment. Also covers blowouts, fires, explosions and miscellaneous accidents, summary of hydrocarbon spills, oil lost, and royalty income.

U.S. Library of Congress. Congressional Research Service. Science Policy Research Division. ENERGY FACTS. Washington, D.C.: Government Printing Office, 1973.

> Contains a comprehensive selection of U.S. and foreign energy statistics.

_____. ENERGY FACT II. Washington, D.C.: Government Printing Office, 1975.

> This is an updated and revised version of the ENERGY FACTS published in 1973 (see above), containing statistical tables and graphs grouped by resources, production, consumption, research and development, and other categories. There is an excellent glossary and a subject index as well as conversion factors and a list of common energy abbreviations.

_____. PROJECT INDEPENDENCE: U.S. AND WORLD ENERGY OUTLOOK THROUGH 1990. Washington, D.C.: Government Printing Office, 1977.

U.S. National Committee of the World Energy Conference (1974). SURVEY OF ENERGY RESOURCES IN ENGLISH AND FRENCH. London: World Energy Conference Central Office; Washington, D.C.: McGregor and Werner, 1974.

> Energy resources studied include coal, crude oil, natural gas, natural gas liquids, oil shale, bituminous sands, hydraulic resources, nuclear fuels (uranium and thorium), tidal energy, geothermal energy, solar energy, wind energy, and ocean thermal gradients. Information is provided on location, geology, technology, uses, historical development, consumption, characteristics, and production. Environmental considerations are discussed for each energy source.

U.S. National Petroleum Council. Committee on the U.S. Energy Outlook. THE UNITED STATES ENERGY OUTLOOK. Washington, D.C.: Government Printing Office, 1971.

> A prior report by the committee, entitled AN INITIAL APPRAISAL 1971-85 (2 vols.), is also available. Both the reports form a comprehensive study of the nation's energy outlook. They con-

tain statistical information on energy supply and demand balance; availability of domestic oil, gas, coal, oil shale, tar sands, nuclear energy, hydroelectric and geothermal water, and fuels for electricity; foreign oil and gas availability, and oil and gas logistics and imports. The trends beyond 1985 are also explored.

U.S. National Science Foundation. INFORMATION ON INTERNATIONAL RESEARCH AND DEVELOPMENT ACTIVITIES IN THE FIELD OF ENERGY. Washington, D.C.: Government Printing Office, 1976.

Data on over 1,700 ongoing and recently completed energy research projects conducted in over thirty countries of the world.

Utah. University of. Bureau of Economic and Business Research. 1976 STATISTICAL ABSTRACT OF UTAH. Salt Lake City: 1976.

Valley National Bank. ARIZONA STATISTICAL REVIEW. Phoenix, Ariz.: 1945-- . Annual.

Vermont. Department of Budget and Management. VERMONT FACTS AND FIGURES. 3d ed. Montpelier: 1975.

Virginia. University of. Thomas Jefferson Center for Political Economy. STATISTICAL ABSTRACT OF VIRIGINIA. Charlottesville: vol. 1, 1966; vol. 2, 1970.

Washington. State Office of Program Planning and Fiscal Management. STATE OF WASHINGTON POCKET DATA BOOK. Olympia: 1975.

Washington. State Research Council. THE RESEARCH COUNCIL'S HAND-BOOK 4th ed. Olympia: 1973. Annual supplements.

WEEKLY ENERGY REPORT. Washington, D.C.: 1973-- . Loose-leaf.

Contains, among other things, a statistical summary of trends in the oil and gas, synthetic fuel, nuclear, and fossil fuel industry as well as electricity transmission and transportation. International in scope.

West Virginia Research League, Inc. THE 1975 STATISTICAL HANDBOOK. Charleston: 1975.

Wisconsin. Department of Administration. Bureau of Planning and Budget. Information Systems Unit. WISCONSIN BLUE BOOK. Madison. Biennial.

_____. WISCONSIN STATISTICAL ABSTRACT. 3d ed. Madison: 1974.

Workshop on Alternative Energy Strategies. ENERGY SUPPLY-DEMAND INTE-
GRATION TO THE YEAR 2000. WAES Techncial Report no. 3. Edited by
Paul S. Basile. Cambridge: MIT Press, 1977.

> Inventory of available energy assets including oil and natural
> gas, coal and nuclear fuels, as well as other fossil and renew-
> able fuels, solar, geothermal, hydroelectric, and waste. The
> WAES project director is Carrol L. Wilson.

World Oil. WORLD OIL MARKET DATA. Houston: 1959-- . Annual.

> Contains data on production, consumption, trade, and prices in
> the world oil market.

WORLD TRADE ANNUAL. 5 vols. New York: Walker & Co., 1963-- .

> Published for the United Nations, this compendium contains data
> on energy-related commodity trade by country arranged by stan-
> dard international trade classification.

Wyoming. University of. Division of Business and Economic Research. WYOM-
ING DATA BOOK. Laramie: 1972.

ENERGY AND TRANSPORTATION

Transportation is one of the industries severely feeling the pinch of the current
energy crisis. Many books and serial publications have appeared on the market
containing statistical information on areas such as fuel consumption and demand
by transportation sector, passenger and freight traffic, and the like. This section contains an annotated listing of over a hundred sources devoted exclusively
to a statistical analysis of transportation and energy.

Aerospace Industries Association of America. AEROSPACE FACTS AND FIG-
URES. New York: Published for the author by Aviation Week and Space
Technology, 1956-- .

> A statistical compendium of the aerospace industry data.

AIRLINE STATISTICAL ANNUAL. Washington, D.C.: Ziff-Davis Publishing
Co., 1973-- .

> Operating statistics on all U.S. airlines, taken from CAB reports.
> Tables and charts cover total industry figures, trunk lines, local
> services, and air cargo lines.

Air Transport Association of America. AIR TRANSPORT FACTS AND FIGURES.
Washington, D.C.: 1940-- . Annual.

> This is a statistical review of the air transport industry. Contains
> some energy-related data.

231

_____. QUARTERLY REVIEW OF AIRLINE TRAFFIC AND FINANCIAL DATA. Washington, D.C.: 1976-- .

Supplements the data given in the annual AIR TRANSPORT FACTS AND FIGURES (see above).

AMERICAN BUREAU OF SHIPPING. RECORD. New York: 1908-- . Annual, with semimonthly supplements.

Data on all American flag ships and many ships of foreign countries.

AMERICAN CAR PRICES MAGAZINE. La Crescenta, Calif.: JEK Publishing, 1976-- . Bimonthly.

Data on road test of all types of cars, including air pollution emission information.

American Public Transit Association. TRANSIT FACTBOOK. Washington, D.C.: 1974-- . Annual.

Contains summary of trends in urban mass transportation, including energy consumption by transit vehicles, energy requirements of passenger transportation modes, and energy comparison of urban transportation modes.

American Taxicab Association. ATA DATABOOK. Chicago. Annual.

Statistical analysis of the taxicab industry.

American Trucking Association. AMERICAN TRUCKING TRENDS. Washington, D.C. Annual.

Statistical data on the industry including energy-related information such as fuel expenses per mile.

_____. FINANCIAL ANALYSIS OF THE MOTOR CARRIER INDUSTRY. Washington, D.C. Annual.

Review of the financial results of class I and class II motor carriers of general commodities, class I special commodity carriers, and publicly held carriers. It gives the composite balance sheet and income statement of the industry, together with select financial and operating ratios.

American Waterways Operators, Inc. INLAND WATERBORNE COMMERCE STATISTICS. Arlington, Va.: 1959-- . Annual.

Published for the Barge and Towing Industry Association, this report contains statistics on inland waterborne commerce in terms of total tonnage moved and total miles of service performed, nationally and by individual waterways; total number of vessels in operation and a comparison of transport services performed by various modes; data on commercially navigable waterways; number of towing vessels and traffic and commodity flows on each specific waterway.

Association of American Railroads. Economics and Finance Department. YEARBOOK OF RAILROAD FACTS. Washington, D.C.: 1965-- . Annual.

Review of the railroad industry containing significant energy-related statistical data such as fuel consumption and passenger and freight traffic by type of commodity like coal.

Association of Oil Pipelines. SHIFTS IN PETROLEUM TRANSPORTATION. Washington, D.C.: 1976-- . Irregular.

Periodical report on the mode of transportation for crude petroleum and petroleum products, the chief modes being pipelines, water carriers, motor carriers, and railroads.

AUTOMOTIVE NEWS ALMANAC. Detroit: Crain Communications, 1933-- . Annual.

Annual statistical issue of the weekly AUTOMOTIVE NEWS. It reviews the entire industry from the point of view of the following: new car and truck production and registration, prices, specifications, and equipment.

Ayres, Robert U. WORLDWIDE TRANSPORTATION ENERGY DEMAND FORECAST: 1975-2000. Springfield, Va.: National Technical Information Service, 1978. 95 p.

COMMERCIAL AIRCRAFT FLEETS. Formerly LOCKHEED COMMERCIAL AIRCRAFT FLEETS. Miami, Fla.: Avmark, 1976-- . Semiannual.

Statistical information on aircrafts used in commercial aviation.

French, Alexander. ENERGY AND FREIGHT MOVEMENTS. Washington, D.C.: U.S. Department of Transportation, 1976.

Data on truck fuel consumption and performance for use in comparing fuel efficiency of trucks with other freight transportation modes.

_____. HIGHWAY TRANSPORT AND ENERGY PROBLEMS IN EUROPE. By Alexander French and Lyle D. Wylie. Washington, D.C.: U.S. Department of Transportation, 1975.

Data on postembargo motor vehicle use and energy consumption in Belgium, France, West Germany, Sweden, and Great Britain. Includes conservation measures, accidents, user tax revenues, and highway expenditures.

Inter Avia. WORLD DIRECTORY OF AVIATION AND ASTRONAUTICS. Geneva: 1946-- . Annual.

Data on operations, scheduled and nonscheduled; administrations, organizations, and institutes; parent and financial organizations; aircraft manufacturers; equipment manufacturers; airports; ground services and equipment; and aviation services. Also published INTERAVIA DATA which reports frequent surveys of the aircraft industry.

International Air Transport Association. WORLD AIR TRANSPORT STATISTICS. Montreal: 1957-- . Annual.

Financial and related data on world scheduled airlines financial results.

JANE'S WORLD RAILWAYS AND RAPID TRANSIT SYSTEMS. London: Jane's Yearbooks; New York: Franklin Watts, distributor, 1956-- . Annual.

Information on the manufacturers of worldwide locomotives and rolling stock. Includes data on the following: freight and passenger traffic, financial situation, container and unit load services, and motive power.

Kirby, Ronald F., et al. PARA-TRANSIT: NEGLECTED OPTIONS FOR UR-BAN MOBILITY. Springfield, Va.: National Technical Information Service, 1974.

Data on para-transit services and their potential for meeting urban travel demands and energy consumption.

MOODY'S TRANSPORTATION MANUAL. New York: Moody's Investor Service, 1928-- . Annual.

Provides financial and related data on the entire transportation industry, including oil pipeline companies.

Motor Vehicle Manufacturers Association. AUTOMOBILE FACTS AND FIGURES. Detroit: 1927-- . Annual.

Published since 1919, this report presents a statistical picture of the automobile industry in the United States. See also the companion publication, below.

_____. MOTOR TRUCK FACTS. Detroit: 1927-- . Annual.

A companion publication to the MVMA's AUTOMOBILE FACTS
AND FIGURES (see above). It provides data on the motor truck
industry.

National Association of Motor Bus Owners. ANNUAL REPORT. Washington,
D.C.: 1927-- .

Provides operating revenues, expenses, and income and balance
sheet data on the nation's 950 intercity bus companies. Reports
growth of package express revenues, impact of competition from
AMTRAK, travel miles, passengers, and employment.

National Association of Motor Bus Owners. GET ON TO SOMETHING GREAT.
INTERCITY AND SUBURBAN BUS INDUSTRY. Washington, D.C.: 1927-- .
Annual.

Formerly issued under title BUS FACTS, this is a review of the
intercity and suburban bus industry in the United States. The
data are compiled from reports presented to the Interstate Commerce Commission and NAMBO's own estimates. Includes statistical data on the industry.

PANAMA CANAL REVIEW. Balboa Heights, Canal Zone: Panama Canal Co.,
1950-- . Semiannual.

Report on shipping statistics including transits and tons of cargo
passing through, given by flag of vessel, direction, trade route,
commodity, and whether U.S. government or commercial.

R.L. POLK AND CO. INFORMATION SYSTEM. Detroit.

A major data collection source on the motor vehicle industry.
Information includes the following: motor vehicle production,
cars and trucks in use by age, new motorcycle registrations by
state.

St. Lawrence Seaway Development Corp. TRAFFIC REPORT OF THE ST.
LAWRENCE SEAWAY. Washington, D.C.: 1959-- . Annual.

Data on number of ships, cargo and revenue passenger volumes,
total revenues, and lockages.

Simat, Helliesen and Eichner, Inc. FORECAST OF AIR TRAFFIC DEMAND
AND ACTIVITY LEVELS TO THE YEAR 2000. Published for Aviation Advisory
Commission. Publication no. PB 216 252. Springfield, Va.: National Tech-
nical Information Service, 1972.

Society of Automotive Engineers. PASSENGER CAR FUEL ECONOMY TRENDS
THROUGH 1976. SAE paper no. 750957. New York: 1975.

SUN OIL CO. ANALYSIS OF WORLD TANKSHIP FLEET. St. Davids, Pa.: 1976-- . Annual.

> Data on oil transportation. It includes data on U.S. tank ship fleet, world tank ship fleet, world tank ship fleet by flag and year of construction, and world tank ships under construction or on order.

Transportation Association of America. TRANSPORTATION FACTS AND TRENDS. Washington, D.C.: 1964-- . Annual.

> This provides a quick reference source of transportation data including the energy needs of the transportation industry, such as transportation demand versus total demand for petroleum.

TRAVEL TRENDS IN THE UNITED STATES AND CANADA. 4th ed. Boulder: Business Research Division, Graduate School of Business Administration, University of Colorado, in cooperation with the Travel Research Association, 1975.

> Contains travel and tourist data on all of the states and provinces and covers such subjects as tourists' visits to recreation areas, tourist expenditures, indicators of economic impact, transportation of out-of-state visitors, tourism advertising, the family vacation markets, passport statistics, and foreign visitors' arrivals.

Trinc Transportation Consultants. TRINC'S RED BOOK OF THE TRUCKING INDUSTRY. Washington, D.C.: 1946-- . Quarterly.

> Selected revenue, income, expense, and traffic statistics for nearly 3,000 class I and class II motor carriers of property, drawn from quarterly reports filed with the Interstate Commerce Commission.

United Nations. Economic Commission for Europe. ANNUAL BULLETIN OF TRANSPORT STATISTICS FOR EUROPE. Geneva: 1950-- .

> Contains data on international goods transport by various modes of transport, including oil pipelines and fuel consumption by road vehicles.

U.S. Bureau of the Census. CENSUS OF TRANSPORTATION. Washington, D.C.: Government Printing Office, 1972-- .

> This census provides data on personal travel, the characteristics and use of trucks, and the shipments of commodities by manufacturers. It consists of the following: The National Travel Survey, which describes the characteristics of "Americans on the Move," the Commodity Transportation Survey, and the Truck Inventory and Use Survey.

_____. FOREIGN TRADE REPORTS. Washington, D.C.: Government Printing Office, 1957-- . Monthly and annual.

These reports of varying frequency give the volume, direction
and value of imports and exports of energy products by method
of transportation, that is, whether by sea or air.

_____. JOURNEY TO WORK. Publication no. PC (2)-6D. Washington,
D.C.: Government Printing Office, 1972.

Data on means of transportation to work for urban and rural population, those living in SMSAs and central cities, and those living outside. The categories available are private automobile
(as a driver or passenger), bus or street car, railroad, taxicab,
subway or elevator train, and walking.

U.S. Center for Naval Analysis. FORECAST OF AIR TRAVEL DEMAND AND
AIRPORT AND AIRWAY USE IN 1980. Publication no. AD 720 732. Springfield, Va.: National Technical Information Service, 1971.

U.S. Civil Aeronautics Board. AIR CARRIER TRAFFIC STATISTICS. Washington, D.C.: Government Printing Office, 1955-- . Monthly.

Report on the traffic, capacity, and performance of certified and
supplemental air carriers. Also includes nonrevenue passenger
miles and U.S. mail revenues.

_____. COMMUTER AIR CARRIER TRAFFIC STATISTICS. Washington, D.C.:
Government Printing Office, 1970-- . Semiannual.

Statistical report on passenger, cargo, and mail air traffic of
individual commuter air carriers, number of points served, and
types of aircraft in operation.

_____. FUEL CONSUMPTION AND COST. Washington, D.C.: Government
Printing Office, 1976-- . Monthly.

Data on fuel consumption and costs of individual trunk, local
service, and supplemental air carriers. Details include fuel
consumption and costs, average price per gallon, and percentage
change in price per gallon.

_____. HANDBOOK OF AIRLINE STATISTICS. Washington, D.C.: 1973.

A statistical analysis of the U.S. certified air carriers and supplemental air carriers. It contains comparative traffic and financial
data including consumption of various types of fuel.

_____. SUPPLEMENT TO HANDBOOK OF AIRLINE STATISTICS. Washington,
D.C.: Government Printing Office, 1976-- . Biennial.

Contains statistical description of individual U.S. airlines and

comparative energy data like consumption of aviation gasoline
and passenger traffic.

U.S. Congress. House. Public Works and Transportation Committee. NA-
TIONAL HIGHWAY NEEDS REPORT. Washington, D.C.: Government Printing Office, 1968-- . Biennial.

Data on highway mileage, capacity, improvement needs, federal
or nonfederal aid, forecasts of highway travel, fuel consumption
and rates, needs of individual urban and nonurban areas, and
costs of achieving a standard level.

U.S. Congress. Office of Technology Assessment. ENERGY, THE ECONOMY,
AND MASS TRANSIT. Summary Report. Washington, D.C.: 1975.

Report prepared by system Design Concepts, Inc. Contains data
on the effects of energy shortage on mass transit ridership,
direct and multiplier effects of transit construction on local and
national employment, transit energy consumption compared with
other modes, and potential governmental actions to increase
transit ridership and conserve energy.

U.S. Department of Commerce. Business and Defense Services Administration.
WORLD MOTOR VEHICLE PRODUCTION AND REGISTRATION. Washington,
D.C.: 1958-- . Annual.

Data on the International Automobile industry.

U.S. Department of Commerce. Maritime Administration. A STATISTICAL
ANALYSIS OF THE WORLD'S MERCHANT FLEETS. Washington, D.C.:
1956-- . Annual.

Worldwide analysis showing the age, size, speed, and draft by
frequency groupings of the world fleets having 100 or more
vessels of 1,000 gross tons or over. Public and private vessels
of the following type: combination passenger and cargo carriers;
freighters; bulk carriers, including bulk oil and ore/oil carriers
and tankers. Both the gross tonnage and dead weight tonnage are
given for each type of vessel.

_____. VESSEL INVENTORY REPORT. Washington, D.C.: Government
Printing Office, 1967-- . Semiannual.

Includes merchant type ships of 1,000 gross tons and over, under
the U.S. registry. There is an alphabetical listing of all such
vessels in the U.S. merchant fleet, whether privately owned or
Maritime Administration owned, showing for each ship the name
of the owner and/or operator, vessel type, design, and deadweight tonnage.

U.S. Department of Transportation. ANALYSIS AND MANAGEMENT OF A PIPELINE SAFETY INFORMATION SYSTEM. Publication no. DOT-TST-75-47. Springfield, Va.: National Technical Service, 1975.

> Data on number, rate, and causes of gas pipeline system leaks and accidents, compiled from annual gas pipeline operator reports. Shows leak rates and distribution by cause, type of pipe material, pipe diameter, part involved, system age, use of coating or cathodic corrosion protection, and region.

_____. ANNUAL REPORT OF THE SECRETARY OF TRANSPORTATION ON THE ADMINISTRATION OF THE NATURAL GAS PIPELINE SAFETY ACT OF 1968. Washington, D.C.: 1969-- .

> Data on pipeline failures and resulting fatalities and injuries, causes of failure, and federal assistance to state agencies for gas pipeline safety programs.

_____. CHARACTERISTICS OF URBAN TRANSPORTATION SYSTEMS: HANDBOOK OF TRANSPORTATION PLANNERS. Washington, D.C.: Government Printing Office, 1974.

> Contains operating and cost information on three modes of urban transportation, namely, rapid rail and commuter rail transit, local bus and bus rapid transit, and automobile highway system. Data include speed, fuel consumption, pollutant emissions, capital costs, and accident frequencies.

_____. ENERGY PRIMER: SELECTED TRANSPORTATION TOPICS. Washington, D.C.: 1975.

> Data on transportation energy consumption, future needs, transportation energy demand, conservation measures and policies, fuel consumption, and transportation costs for interurban and intraurban passenger travel, classified by mode and by freight and passenger traffic energy consumption and needs for the period 1950-2000.

_____. ENERGY USE AND OTHER COMPARISONS BETWEEN DIESEL AND GASOLINE TRUCKS. By Kenneth M. Jacobs. Springfield, Va.: National Technical Information Service, 1977. 107 p.

> Compares fuel consumption and operating costs for gasoline and diesel powered trucks.

_____. FUEL CONSUMPTION OF TRACTOR-TRAILER TRUCKS AS AFFECTED BY SPEED LIMIT AND PAYLOAD WEIGHT. By Anthony J. Broderick. Springfield, Va.: National Technical Information Service, 1975.

> Results of a 1974 test comparing fuel consumption of two tractor trailer trucks, carrying payload weights of 28,000 and 42,000 pounds respectively, at three speeds, over two types of terrain.

_____. NATIONAL TRANSPORTATION REPORT. Washington, D.C.: 1972-- . Annual.

Assesses the American transportation system and discusses alternatives for improving transportation performance in the future. Contains data on transportation plans and priorities on federal, state, and local levels.

_____. PROFILES OF PUBLIC TRANSPORTATION PLANS AND PROGRAMS. Washington, D.C.: Government Printing Office, 1975-- . Annual.

Issued as an appendix to the annual National Transportation Report, this presents a detailed description and analysis of the public transportation data submitted by 52 largest urbanized areas, with a population of half a million or more.

_____. STATEMENT OF NATIONAL TRANSPORTATION POLICY. Washington, D.C.: Government Printing Office, 1972-- . Annual.

Contains data on national transportation policy recommendations by mode, financing of federal programs and subsidies by type, and transportation system performance measures.

_____. STUDY OF POTENTIAL FOR MOTOR VEHICLE FUEL ECONOMY IMPROVEMENT. PANEL REPORTS. Washington, D.C.: 1975.

Series of seven reports prepared by special joint DOT and EPA Panels to investigate the implications of a proposed national fuel economy improvement standard of 20 percent for new motor vehicles by 1980.

_____. SUMMARY OF OPPORTUNITIES TO CONSERVE TRANSPORTATION ENERGY. By John Pollard et al. Springfield, Va.: National Technical Information Service, 1975.

Presents 1972 base year data on transportation energy consumption and energy efficiency of each transportation mode. Includes potential energy savings from conservation measures, including improvements in vehicle design and operating efficiency, increased load factors, reductions in service or use, and modal shifts.

U.S. Department of Transportation. Office of Secretary. SUMMARY OF NATIONAL TRANSPORTATION STATISTICS. Washington, D.C.: Government Printing Office, 1971-- . Annual.

Compendium of selected national transportation statistics. It includes data on cost, inventory, and performance of oil pipeline industries and fuel consumption by mode of transportation.

U.S. Department of Transportation. Transportation Systems Center. SMALL

CITY TRANSIT. 15 vols. Publication nos. PB 251 501 to PB 251 515. Springfield, Va.: National Technical Information Service, 1976.

> Report sponsored by the Urban Mass Transportation Administration. Contains data on the following: small city transit characteristics, free fare, student operated transit in a university community, dial-a-ride project, private subscription bus service, public transit, low subsidies transit, and transportation costs.

U.S. Energy Research and Development Administration. TRANSPORTATION ENERGY CONSERVATION DATA BOOK. By D.B. Shonka et al. Washington, D.C.: Government Printing Office, 1976-- . Semiannual.

> Compilation of statistics relating to energy use in the transportation sector. Provides detailed data for 1950 through the mid-1970s on vehicle mileage, sales, and energy consumption; petroleum production and imports; and demographic and economic determinants of vehicle use.

U.S. Environmental Protection Agency. FACTORS AFFECTING AUTOMOTIVE FUEL ECONOMY. Washington, D.C.: 1976.

> Report on automobile fuel economy, based on extensive EPA tests of foreign and domestic cars, with and without emission controls. Includes data on vehicle and engine design factors, and operation and use factors affecting fuel economy trends.

_____. GAS MILEAGE GUIDE FOR NEW CAR BUYERS. Washington, D.C.: Government Printing Office, 1976-- . Annual.

> Fuel economy results on new model cars and trucks tested by EPA, arranged by manufacturer and car line, with engine size, number of cylinders, carburetor type, whether catalyst used, and fuel in miles per gallon for city and highway driving.

U.S. Federal Aviation Administration. AVIATION FORECASTS. Springfield, Va.: National Technical Information Service, 1968-- . Annual.

> Fiscal year forecasts of indicators of aviation activity and traffic for general aviation, air carriers, and military.

_____. CENSUS OF U.S. CIVIL AIRCRAFT. Washington, D.C.: Government Printing Office, 1967-- . Annual.

> Contains a count of all registered civil aircraft in the United States. Both current and historical data are provided.

_____. FAA AIR TRAFFIC ACTIVITY. Washington, D.C.: Government Printing Office, 1956-- . Annual.

> Analysis of air traffic activity at FAA-operated airport traffic

control towers, air force and navy radar approach control facilities, and resident air traffic specialists locations.

_____. FAA STATISTICAL HANDBOOK OF CIVIL AVIATION: Washington, D.C.: 1944-- . Annual.

The handbook serves as a convenient source for historical data on the national air space system, airports, and airport activity.

_____. NATIONAL AVIATION SYSTEM PLAN. Washington, D.C.: Government Printing Office, 1970-- . Annual.

Forecasts of FAA activity levels, facilities, equipment, and manpower.

_____. PROFILES OF SCHEDULED AIR CARRIER PASSENGER TRAFFIC: TOP U.S. AIRPORTS. Washington, D.C.: Government Printing Office, 1974-- . Annual.

Report on the number of passengers arriving and departing on scheduled commercial planes at the top 100 U.S. airports by the hour.

_____. SELECTED STATISTICS OF PUBLISHED DOMESTIC AIR CARRIER SCHEDULES. Washington, D.C.: Government Printing Office, 1973-- .

Four-part report on air carrier domestic scheduled and estimated flight load factors, by air carrier, airport, equipment type, flight stage length, origins and departures and flown seat miles.

_____. TERMINAL AREA FORECAST 1977-87. Springfield, Va.: National Technical Information Service, 1974-- . Annual.

This is an annual report presenting forecasts of aviation activity including passenger traffic and operations levels.

U.S. Federal Energy Administration. Project Independence. ENERGY CONSERVATION IN TRANSPORTATION SECTORS. Washington, D.C.: Government Printing Office, 1974.

This report was prepared by Jack Faucett and Associates, Inc., under the direction of the Council of Environmental Quality. It covers energy conservation and anlyzes the impact of increased fuel prices on passenger and freight transportation and energy conservation potential in the transport sector, through 1980, 1985, and 1990.

U.S. Federal Highway Administration. ANNUAL MILES OF AUTOMOBILE TRAVEL. Nationwide Personal Transportation Study, no. 2. Washington, D.C.: Government Printing Office, 1972.

Report on automobile use by household, including data on number of cars in the household, age of the automobile by year model, cars purchased new or used, annual income of the household, occupation of the principal operator of the automobile, and place of residence of principal operator by incorporated places and incorporated areas.

_____. AUTOMOBILE OCCUPANCY. Nationwide Personal Transportation Study Report, no. 1. Washington, D.C.: Government Printing Office, 1972.

Data on auto trips, number of occupants on each trip, passenger miles, vehicle miles, vehicle occupancy rates, and the purposes of travel, like earning a living, family business, educational, civic, religious, social, and recreational.

_____. AUTOMOBILE OWNERSHIP. Nationwide Personal Transportation Study Report, no. 11. Washington, D.C.: Government Printing Office, 1974.

Presents data on car ownership, place of principal residence, household income, vehicle trips, and vehicle miles.

_____. AVAILABILITY OF PUBLIC TRANSPORTATION AND SHOPPING CHARACTERISTICS OF SMSA HOUSEHOLDS. Washington, D.C.: Government Printing Office, 1974.

Nationwide personal transportation study on the availability of public transportation to the main business district of the central city for households located in SMSAs, and information on shopping characteristics and income of the household by race of household.

_____. AVERAGE DAILY TRAFFIC ON RURAL MILEAGE OF THE INTERSTATE SYSTEM TRAVELLED-WAY. Washington, D.C.: Government Printing Office, 1976-- . Annual.

Average daily vehicle miles of traffic on the national interstate traffic system by rural and urban location, access control, and by state and selected major city.

_____. COST OF OWNING AND OPERATING AN AUTOMOBILE. Washington, D.C.: 1960-- . Biennial.

Periodical report on total cost and cents-per-mile cost of operating an automobile, including expenditure on oil and gasoline.

_____. HIGHWAY STATISTICS. Washington, D.C.: Government Printing Office, 1945-- . Annual.

Statistical data on the ownership and operation of motor vehicles, receipts and expenditures for highways by public agencies, and

the characteristics of the mileage of public highways and roads.
Includes an analysis of motor fuel consumption, state motor-fuel
tax receipts, and disposition.

_____. HIGHWAY TRAVEL FORECASTS. Springfield, Va.: National Techni-
cal Information Service, 1974.

Forecasts of number of drivers, miles per driver, and rates of increase of the number of automobiles and gas consumption for
seven age groups. These are given under varying assumptions
regarding fuel efficiency and gasoline availability. Historical
data from 1940-74 is available on the following: traffic volume
trends compared to other modes, and characteristics of highway
travelers like age, sex, income, and residency.

_____. HOME TO WORK TRIPS AND TRAVEL. Washington, D.C.: Govern-
ment Printing Office, 1973.

Data on the characteristics of home-to-work travel, choice of
transportation mode involved, and the importance of automobiles
in such travel. The data are analyzed under different variables
such as distance, time of travel, residence, population groups,
occupation, age, and hours of the day.

_____. HOUSEHOLD TRAVEL IN THE UNITED STATES. Nationwide Personal
Transportation Study Report, no. 7. Washington, D.C.: Government Printing
Office, 1972.

The highlights of this report include the following: passenger
car travel per household, daily trip making by major purpose,
average number and length of trips, and income characteristics
of people making trips.

_____. IMMEDIATE IMPACT OF GASOLINE SHORTAGES ON URBAN TRAVEL
BEHAVIOR. By Robert L. Peskin et al. Washington, D.C.: 1975.

Data on the relationship between the price and availability of
gasoline and the use of automobiles or alternate transportation
modes for various purposes.

_____. MODE OF TRANSPORTATION AND PERSONAL CHARACTERISTICS
OF TRIP MAKERS. Nationwide Personal Transportation Study Report, no. 9.
Washington, D.C.: Government Printing Office, 1973.

This report presents travel distribution by age, sex, race, and
place of residence. It contains data on the following: mode
of transportation by age group, annual auto trips for males and
females, and trips by public transportation.

_____. PURPOSE OF AUTOMOBILE TRIPS AND TRAVEL. Nationwide Personal Transportation Study Report, no. 10. Washington, D.C.: Government Printing Office, 1974.

> Data on four major purposes of automobile trips; (1) earning a living, (2) family business, (3) educational, and (4) civic, religious, social, and recreational activities. Trips are analyzed in terms of trip length, age of driver, occupation, household income, hour of the day the trip started, day of the week, season of the year, and the number of occupants per trip.

_____. TRANSPORTATION CHARACTERISTICS OF SCHOOL CHILDREN. Nationwide Personal Transportation Study Report, no. 4. Washington, D.C.: Government Printing Office, 1972.

> Report on travel patterns to school of students between 5 and 18, kindergarten through grade 12. For each grade level, home-to-school travel by various modes of transportation is analyzed in terms of distance to school and time from home to school. Some data are given on students attending public and private schools.

U.S. Federal Highway Administration. Highway Statistics Division. MOTOR FUEL CONSUMPTION DATA. Washington, D.C.: FHA Public Affairs Office, 1935-- .

> Frequent releases containing data on motor fuel consumption by state and type of use, compiled from state tax records.

_____. MOTOR GASOLINE SALES. Washington, D.C.: FHA Public Affairs Office, 1972-- . Monthly.

> Contains information on gross motor gasoline sales for the most recent quarter.

U.S. Interstate Commerce Commission. FINANCIAL AND OPERATING STATISTICS OF CLASS I MOTOR CARRIERS OF PASSENGERS. Washington, D.C.: Government Printing Office, 1971-- . Semiannual.

> This is a report on passenger carrier operating revenues, expenses, income, highway and vehicle mileages, and number of passengers by type of route. Coverage includes intercity and local or suburban carriers. The report was formerly issued under the title REVENUES, EXPENSES, OTHER INCOME AND STATISTICS OF CLASS I MOTOR CARRIERS OF PASSENGERS.

_____. FINANCIAL AND OPERATING STATISTICS OF CLASS I MOTOR CARRIERS OF PROPERTY. Washington, D.C.: 1971-- . Semiannual.

> Report on total operating revenues and expenses by type, income, and mileage of intercity and local motor carriers of property.

_____. REPORT. Washington, D.C. Annual.

Review of that portion of the transportation industry which falls under regulatory jurisdiction of the ICC, including pipelines. It contains general data on the following: structure, reorganization, labor, abandonments, mergers, passenger service, revenue and rates, safety, construction, and operation.

_____. REVENUES, EXPENSES AND STATISTICS OF THE FREIGHT FOR-WARDERS. Washington, D.C.: Government Printing Office, 1943-- . Semi-annual.

Data on operating revenues, expenses, and income of freight forwarders having gross yearly revenues of at least $1,000,000.

_____. STATISTICS OF RAILWAYS OF THE UNITED STATES. Washington, D.C.: Government Printing Office, 1887-- . Annual.

Historical data on the railroad industry. Superseded by TRANS-PORT STATISTICS OF THE UNITED STATES SINCE 1953.

U.S. National Highway Safety Administration. AUTOMOTIVE FUEL ECON-OMY PROGRAM, ANNUAL REPORT TO THE CONGRESS. Washington, D.C.: Government Printing Office, 1977-- .

Contains charts and tables showing trends in petroleum production and fuel consumption, 1960-85, passenger and nonpassenger auto-mobile standards, average vehicle weight and fuel economy, and proposed nonpassenger automobile fuel economy standards, model years 1980-81.

U.S. Travel Data Center. IMPACT OF TRAVEL ON STATE ECONOMIES, 1974. Washington, D.C.: 1976.

Analyzes the travel industry and how it affects state economies.

URBAN MASS TRANSPORTATION. Special study, no. 82. Cleveland: Predicasts, 1973.

This report analyzes the past, present, and future of mass transit and mass transit equipment industries, in the light of demographic trends, preference patterns, equipment, and fuel costs.

URBAN MASS TRANSPORTATION. Special Study, no. 3-76-122. Cleveland: Predicasts, 1976.

Comprehensive revision of a 1973 study (see above) analyzing the urban mass transportation needs in the United States through 1990. Changing energy costs, conveniences, and regulations affecting mode of transportation preference patterns are discussed in detail.

WARD'S AUTOMOTIVE YEARBOOK. Detroit: Wards Communications, 1938-- .

Data on the following: auto bodies and equipment; factory shipments; imports, registrations; production of cars, trucks, and recreational vehicles; automotive sales; and auto parts market.

WORLD AIRLINE RECORD. Chicago: Roadcap Associates, 1948-- . Annual, with quarterly supplements.

Comprehensive reference on every scheduled airline in the world. Includes data on traffic and equipment.

WORLD AIR TRAFFIC FORECAST. Burbank, Calif.: Lockheed California Co., 1976-- . Annual.

WORLD CARS. Rome: Automobile Club of Italy, 1962-- . Distributed by Herald Books, N.Y. Annual.

Contains data on world automobile production. Statistical information is provided on individual type of automobiles, including engines, transmission, performance, chassis, steering, brakes, electrical equipment, dimension, body types, and accessories.

WORLD COMMERCIAL VEHICLES. World Studies, no. 135. Cleveland: Predicasts, 1977.

This study analyzes commercial vehicle usage, annual demand, production and assembly, and international trade in over forty countries and all regions of the world. Historical data is presented for 1963-75 and forecasts are given for 1980, 1985, 1990.

DIRECTORY OF PUBLISHERS

Aerospace Industries Association of
America
1725, DeSales Street
Washington, D.C. 20036

Air Transport Association of America
1000 Connecticut Avenue N.W.
Washington, D.C. 20036

Alaska. Dept. of Commerce and
Economic Development
Division of Economic Enterprise
Juneau, Ala. 99811

American Association of Petroleum
Geologists
P.O. Box 979
Tulsa, Okla. 74101

American Bureau of Shipping
45, Broad Street
New York, N.Y. 10034

American Enterprise Institute for Public
Policy Research
1150, 17th Street N.W.
Washington, D.C. 20036

American Gas Association
1515 Wilson Blvd.
Arlington, Va. 22209

American Mining Congress
1100 Ring Building
Washington, D.C. 20036

American Nuclear Society
244 E. Ogden Avenue
Hinsdale, Ill. 60521

American Petroleum Institute
1271, Avenue of the Americas
New York, N.Y. 10036

American Public Transit Association
465 L'Enfant Plaza West, S.W.
Washington, D.C. 20024

American Taxicab Association
222 Wisconsin Avenue
Lake Forest, Ill. 60045

American Trucking Association
1616 P. Street, N.W.
Washington, D.C. 20036

American Waterways Operators Inc.
Suite 1100, 1600 Wilson Building
Arlington, Va. 22209

Ann Arbor Science Publishers Inc.
P.O. Box 1425
Ann Arbor, Mich. 48106

Arab Petroleum Research Center
Box 7167
Beirut, Lebanon

Arkansas Almanac Inc.
P.O. Box 191
Little Rock, Ark. 72203

Directory of Publishers

Arnold Bernhard & Co.
5 East 44th Street
New York, N.Y. 10017

Association of American Railroads
1920 L. Street, N.W.
Washington, D.C. 20036

Association of Oil Pipelines
1725 K. Street, N.W.
Washington, D.C. 20006

Atomic Industrial Forum
1747 Pennsylvania Avenue
N.W. Suite 1150
Washington, D.C. 20006

Ballinger Publishing Company
17 Dunster Street
Harvard Square
Cambridge, Mass. 02138

Bankers Trust Company
16 Wall Street
New York, N.Y.

Barks Publications Inc.
400 N. Michigan Avenue
Chicago, Ill. 60601

Battelle Memorial Institute
505 King Avenue
Columbus, Ohio 43215

British Petroleum Company
Britannic House, Moore Lane
London EC279 BU, England

Brookhaven National Laboratory
Upton, New York 11973

Building Owners and Managers
   Association
1221 Massachusettes Avenue
Washington, D.C. 20005

Business Publishers Inc.
P.O. Box 1067
Silver Spring, Md. 20910

Butane-Propane News Inc.
735 W. Duarte Road
Arcadia, Calif. 91006

C.A. Turner & Associates Inc.
P.O. Box F
Bloomingdale, Ill. 60108

California Dept. of Finance
State Capital, Room 1145
Sacramento, Calif. 95814

Cameron Engineers
1315, S. Clarkson
Denver, Colo.

Canadian Petroleum Association
Calgary, Alberta, Canada

Capital Energy Letter Inc.
National Press Building
Washington, D.C. 20045

Centerline Corp.
401 South 36th Street
Phoenix, Ariz. 85034

Charter House Group
1, Paternoster Row
St. Pauls, London EC4P$HP
England

Chase Manhattan Bank
Energy Economics Dept.
1, Chase Manhattan Plaza
New York, N.Y. 10015

Chemical Specialties Manufacturers
   Association
50 East 41st Street
New York, N.Y. 10017

Colorado Geological Survey
1313 Sherman Street
Room 715
Denver, Colo. 80203

Columbia University Press
136 S. Broadway
Irvington on Hudson
New York, N.Y. 10533

Commodity Research Bureau Inc.
One Liberty Plaza
New York, N.Y. 10006

The Conference Board Inc.
845, 3rd Avenue
New York, N.Y. 10022

Connecticut. Dept. of Commerce.
  Technical Services Division
210 Washington Street
Hartford, Conn. 06106

Corpus Publishers Services Ltd.
6, Crescent Road
Toronto, Ontario, Canada   M4 W1P1

Crain Communications
965, E. Jefferson
Detroit, Mich. 48207

Dallas Morning News
A.H. Belo Corporation
Communications Center
Dallas, Tex. 75222

DeGolyer and MacNaughten
5625 Daniels Avenue
Dallas, Tex. 75206

Delaware. State Planning Office
Thomas Collins Building
Dover, Del. 19901

Economist Intelligence Unit Ltd.
Spencer House
27 St. James Place
London SW' A 1NT England

Edison Electric Institute
90 Park Avenue
New York, N.Y. 10016

Electric Power Research Institute
Palo Alto, Calif.

Electric Letter
John Turrell
Rt. 2, Mt. Vernon, Ill.

Electric Vehicle Council
Box 533
Westport, Conn. 06880

Energy Communications
800 Davis Building
Dallas, Tex. 75202

Environment Information Center
124 E. 39th Street
New York, N.Y. 10016

Exxon Company
Box 2180
Houston, Tex. 77001

Financial Post
McLean Hunter Publishers
481, East Univ. Avenue
Toronto 2, Ontario, Canada

Financial Times
Bracken House
Cannon Street
London, EC4, England

Florida Energy Data Center
Tallahassee, Fla. 32304

Future Requirements Agency
Denver Research Institute
University of Denver
2199 S. University Blvd.
Denver, Colo.

Gas Appliance Manufacturers Association
1901 N. Forth Myer Dr.
Arlington, Va. 22209

Gower Press
1, Westmead
Farnborough, Hants GH147RU,
  England

Government Institutes
4733 Bethesda Avenue, N.W.
Washington, D.C. 20014

Graham & Trotman Ltd.
20 Fouberts Place
Regent Street
London W1V1HH, England

Great Britain. Dept. of Energy
Information Division
Room 1674, Thames House South,
   Millbank
London SW1P4QJ, England

Gulf Publishing Company
3301 Allen Parkway
Houston, Tex. 77001

Harcourt Brade Jovanovitch
757, Third Avenue
New York, N.Y. 10017

Harper & Row Publishers
49 E. 33d Street
New York, N.Y. 10016

Illinois. State Geological Survey
139 Natural Resources Building
Urbana, Ill. 61801

Independent Petroleum Association of
   America
1101, 16th Street, N.W.
Washington, D.C. 20036

Indiana. Geological Survey
Bloomington, Ind. 47401

Indiana. State Planning Services
   Agency
333 State House
Indianapolis, Ind. 46204

Institute for Energy Analysis
Oak Ridge, Tenn. 37830

Institute of Petroleum, Information
   Services Dept.
61, New Cavendish Street
W1, London, England

Interavia Data
P.O. Box 162
CH-1216
Geneva-Cointrin, Switzerland

International Air Transport Association
1155 Mansfield
Montreal, Quebec, Canada

International Atomic Energy Agency
Division of Publications
Kaertner Ring 11, Box 590, A-1011
Vienna, Austria

International Energy Agency
Chateau de la Muette
2, rue Andre Pascal
75755, Paris Cadex 16

International Oil Scouts Association
Box 2121
Austin, Tex. 78767

Interstate Oil Compact Commission
Box 5317
Oklahoma City, Okla. 73152

Iowa. Development Commission.
   Research Division
250 Jewett Building
Des Moines, Iowa 50319

John S. Herold Inc.
35 Mason Street
Greenwich, Conn. 06830

Johns Hopkins Press
Johns Hopkins University
Baltimore, Md. 21218

Kentucky. Dept. of Commerce
Capitol Plaza Tower
Frankfort, Ky. 40601

Kerr and Co. Engineers
Energy Concepts International Inc.
111 Broadway
New York, N.Y. 10006

LaMotte News Bureau
1236 National Press Building
Washington, D.C. 20045

Latin American Center
405 Hilgard Avenue
University of California
Los Angeles, Calif. 90024

Lockheed California Co.
P.O. Box 551
Burbank, Calif. 91520

Louisiana. Dept. of Conservation
State Land and Natural Resources
　Building
Baton Rouge, La. 70804

Louisiana. Division of Natural
　Resources and Energy
State Land and Natural Resources
　Building
Baton Rouge, La. 70804

Louisiana. Geological and Gas Divi-
sion
State Land and Natural Resources
　Building
Baton Rouge, La. 70804

Lundberg Survey Inc.
P.O. Box 3996
North Hollywood, Calif. 91609

M. Palmer Publishing Company
25, W 45
New York, N.Y.

McGraw Hill Book Company
1221 Avenue of the Americas
New York, N.Y. 10036

Maine. State Development Office
State House
Augusta, Me. 04333

Margolis Industrial Services
50 East 41st Street
New York, N.Y. 10017

Maryland. Dept. of Economic and
　Community Development
1748 Forest Drive
Annapolis, Md. 21401

Massachusetts. Dept. of Commerce
　and Development
100 Cambridge Street
Boston, Mass. 02202

Metals Week
1221 Avenue of the Americas
New York, N.Y. 10036

Michigan. Geology Division
Box 30028
Lansing, Mich. 48901

Michigan State University
Graduate School of Business Adminis-
　tration
Division of Research
Ann Arbor, Mich. 48104

Midwest Oil Register
P.O. Box 7248
Tulsa, Okla. 74105

Miller Freeman Publishing Inc.
500 Howard Street
San Francisco, Calif. 94105

Minnesota. Dept. of Economic
　Development. Research Division
480 Cedar Street
St. Paul, Minn. 55101

Minnesota. State Planning Agency.
　Office of Local and Urban Affairs
550 Cedar Street
St. Paul, Minn. 55101

Mississippi. State University
College of Business and Industry
Division of Research
P.O. Box 5288
Mississippi Street, Miss. 39762

Missouri. Energy Agency
1014 Madison Street
Jefferson City, Mo. 65101

Missouri. Public Service Commission
Office of Economic Research
Jefferson City, Mo. 65101

MIT Press
Massachusetts Institute of Technology
Cambridge, Mass. 02139

Montana. Energy Advisory Council
Helena, Mt. 59601

Montana State Division of Research
and Information Systems
State Capitol
Helena, Mt. 59601

Moody's Investors Services Inc.
99 Church Street
New York, N.Y. 10007

Motor Vehicle Manufacturers Association
320 New Center Building
Detroit, Mich. 48202

National Association of Motor Bus
Owners
1025 Connecticut Avenue
Washington, D.C. 20036

National Coal Association
1130, 17th Street, N.W.
Washington, D.C. 20036

National Economic Research
Association
80 Broadway
New York, N.Y. 10004

National LP Gas Association
1301 West 22d
Oakbrook, Ill. 60521

National Science Teachers Association
1201, 16th Street, N.W.
Washington, D.C. 20036

National Technical Information
Service
U.S. Dept. of Commerce
5285 Port Royal Road
Springfield, Va. 22161

Nebraska. Dept. of Economic
Development
Division of Research
301 Centennial Mall
Lincoln, Neb. 68509

Nebraska Oil and Gas Conservation
Commission
P.O. Box 399
Sidney, Neb. 69162

Nevada. Dept. of Economic
Development
Capitol Building
Carson City, Nev. 89710

New Hampshire. Dept. of Resources
and Economic Development
318, State House Annex
Concord, N.H. 03301

New Jersey. Office of Business
Economics
Trenton, N.J. 08625

New York. Divison of Budget.
Office of Statistical Coordination
State Capitol
Albany, N.Y. 12224

New York State Public Service
Commission
Albany, N.Y. 12224

North Carolina. Dept. of Administration. Office of the State Budget
116 W. Jones Street
Raleigh, N.C. 27611

North Carolina. Energy Division
Dept. of Veterans and Military Affairs
215 E. Lane Street
Raleigh, N.C. 27611

North Carolina. Interagency Natural
Gas Task Force
Raleigh, N.C. 27611

North Dakota. Business and Industrial
Development Dept.
523, E. Bismark Avenue
Bismark, N. Dak. 58505

Noyes Data Corporation
Noyes Building
Parkridge, N.J. 07656

Nuclear Assurance Corporation-
24 Executive Park West
Atlanta, Ga. 30329

Oak Ridge Associated Universities
P.O. Box 117
Oak Ridge, Tenn. 37830

Oak Ridge National laboratory
P.O. Box
Oak Ridge, Tenn. 37830

Observer Publishing Company
2420 Wilson Blvd.
Arlington, Va. 22201

O.E.C.D. Publications Center
1750 Pennsylvania Avenue, N.W.
Washington, D.C. 20006

Offshore Rig Data Services
5755 Bonhomme, Suite 400
Houston, Tex. 77036

Ohio. Dept. of Economic and
    Community Development
Office of Population Statistics
30, East Broad Street
Columbus, Ohio 43215

Oil Daily Inc.
850, Third Avenue
New York, N.Y. 10022

Organization of Petroleum Exporting
    Countries
Dr. Karl Lueger-Ring 10
1010 Vienna, Austria

Panama Canal Company
Balboa Heights
Canal Zone

Parra, Ramos and Parra Ltd. of
    Venezuela
Guildgate House, The Terrace
Wokingham, Berkshire
RC 11 1BP, England

Pennsylvania. Dept. of Commerce
    Bureau of Statistics
Research and Planning
419, South Office Building
Harrisburg, Pa. 17120

Petroleum Engineer Publishing Company
P.O. Box 1589
Dallas, Tex. 75221

Petroleum Equipment Suppliers Associa-
    tion
1703, First City National Bank
    Building
Houston, Tex. 77002

Petroleum Industry Research Foundation
122 E. 42d
New York, N.Y. 10017

Petroleum Press Bureau
5, Pemberton Road
Fleet Street
London EC4A3DP, England

Petroleum Publishing Company
P.O. Box 1260
Tulsa, Okla. 74101

Platt's Oilgram Price Service
1221 Avenue of the Americas
New York, N.Y. 10036

Potential Gas Committee
Colorado School of Mines
Golden, Colo. 80401

Predicasts Inc.
200 University Circle Research Center
11001 Cedar Avenue
Cleveland, Ohio 44106

Puerto Rico. Planning Board. Bureau
    of Statistics
Box 9447
Santurce, PR 00908

Quarterly Report of Gas Industry
    Operations Inc.
1515 Wilson Blvd.
Arlington, Va.

Rand Corporation
1700 Main Street
Santa Monica, Calif.   90406

Resource Programs Inc.
521, 5th Avenue
New York, N.Y.   10017

Resources for the Future
1755 Massachusetts Avenue, N.W.
Washington, D.C.   20036

Rhode Island.   Dept. of Economic
Development
1 Weybosset Hill
Providence, R.I.   02903

Roadcap Associates
327 South LaSalle Street
Chicago, Ill.   60604

Robert Morey Associates
P.O. Box 98
Dana Point, Calif.   92629

Rockefeller Foundation
1230 Avenue of the Americas
New York, N.Y.   10036

Sales and Marketing Management Inc.
633 Third Avenue
New York, N.Y.   10017

Sheffer Company
5755 Bonhomme
Houston, Tex.   77036

Societe de Documentation et
d'Analyses Financieras
125 Rue Monmarte
2^e Paris, France

Society of Automotive Engineers
400 Commonwealth Drive
New York, N.Y.

Society of Exploration Geophysicists
Box 3098
Tulsa, Okla.   74101

Solar Energy Industry Association
Suite 800, 1001 Connecticut
Avenue, N.W.
Washington, D.C.   20036

South Carolina.   Budget and Control
Board.   Division of Research and
Statistical Services
205 Wade Hampton Office Building
Columbia, S.C.   57501

South Dakota.   State Planning Bureau
State Capital Building
Pierre, S. Dak.   57501

Standard and Poor Corporation
345 Hudson Street
New York, N.Y.   10014

Stanford University Press
Stanford, Calif.   94305

Statistical Office of the European
Communities
Centre Louvigny
P.O. Box 1907
Luxembourg

Sun Oil Company
Publication Division
Radnor-Chester Road
St. Davids, Pa.   19087

Tennessee.   University Center for
Business and Economic Research
Knoxville, Tenn.   37916

Tetra Tech Inc.
1911 Ft. Myer Dr.
Arlington, Va.   22209

Texas Eastern Transmission Corporation
Southern National Park Building
Houston, Tex.   77002

Texas.   Office of Information Service
Management Science Division
Austin, Tex.   78701

Trends Publishing Inc.
330 National Press Building
Washington, D.C. 20034

Trinc Transporation Consultants
Suite 4200, 485 L'Enfant Plaza S.W.
Washington, D.C. 20024

UNIPUB
Box 433, Murray Hill Station
New York, N.Y. 10016

United Nations, Sales Section
Publishing Service
New York, N.Y. 10017

United Nations, Statistical Office
Room LX-2300
New York, N.Y. 10017

U.S. Atomic Energy Commission
Grand Junction Office
Colorado 81501

U.S. Federal Energy Administration
12th St. and Pennsylvania Avenue,
N.W.
Washington, D.C. 20461

U.S. Government Printing Office
North Capitol and H. Street N.W.
Washington, D.C. 20401

Universal News
1840 Ridgecrest Street
Houston, Tex. 77055

University of Florida
Bureau of Business and Economic
Research
College of Business Administration
Gainesville, Fla. 32611

University of Georgia
College of Business Administration
Division of Research
Athens, Ga. 30601

University of Idaho
Center for Business Development and
Research
Moscow, Idaho 83843

University of Kansas
Institute for Social and Environmental
Studies
Lawrence, Kan. 66045

University of Missouri
Extension Division
Columbia, Mo. 65201

University of New Mexico
Bureau of Business Research
Albuquerque, N.M. 87131

University of New Orleans
Division of Business and Economic
Research
New Orleans, La. 70122

University of Oklahoma
Center for Economic and Management
Research
Norman, Okla. 73069

University of Oregon
Bureau of Business Research
Eugene, Oreg. 97403

University of South Dakota
Business Research Bureau
Vermillion, S. Dak.

University of Utah
Bureau of Economic and Business
Research
Salt Lake City, Utah 84108

University of Virginia
Thomas Jefferson Center for Political
Economy
Charlottesville, Va. 22901

University of Wyoming
Division of Business and Economic
Research
Laramie, Wy. 82070

Utah. Oil and Gas Conservation
Commission
1588 West North Temple
Salt Lake City, Utah 84116

Utilities and Transportation
Commission
Highways-Licenses Building
Olympia, Washington 98504

Valley National Bank
241 North Central Avenue
Phoenix, Ariz. 85001

Vermont. Dept. of Budget and
Management
109 State Street
Montpelier, Vt. 05602

Walker & Co.
720, 5th Avenue
New York, N.Y. 10019

Wards Communications Inc.
28, West Adams
Detroit, Mich. 48226

Wisconsin. Dept. of Administration.
Bureau of Planning and Budget
B114, Wilson Street
State Office Building
Madison, Wis. 53702

Wisconsin. Public Service Commission
422, Hill Farms State Office Building
Madison, Wis. 53702

World Energy Conference
Central Office
5, Bury Street
St. James, London, England

Ziff-Davis Publishing Company
1156, 15th Street, N.W.
Washington, D.C. 20005

PERSONAL AND CORPORATE AUTHOR INDEX

This index includes all editors, authors, corporations and organizations cited in the text. The alphabetization is letter by letter and the references are to page number.

A

Acton, Jean Paul 161
Adelman, Morris A. 177
Aerospace Industries Association of America 231
Air Transportation Association of America 231-32
Alabama University Center for Business and Economic Research 200
Alaska. Dept. of Commerce and Economic Development. Division of Economic Enterprise 200
Allen, Edward 200
American Association of Petroleum Geologists 177
American Bureau of Shipping 232
American Gas Association 149, 169
American Mining Congress 158
American Petroleum Institute 147, 152, 153
American Petroleum Institute. Division of Statistics 170, 177-78
American Public Transit Association 232
American Taxicab Association 232
American Trucking Association 232
American Waterways Operators, Inc. 232
Arab Petroleum Research Center 179

B

Arizona University Division of Economic and Business Research 200
Arkansas Almanac Inc. 200
Arnold Bernhard & Co. 200
Association of American Railroads 233
Association of Oil Pipelines 179, 233
Ayers, Robert U. 233

B

Bacher, Ken 195
Bankers Trust Co. 200
Battelle Memorial Institute 200
Behling, D.J. 200
Bennington, G. 195
Benson, David C. 159
Berman, Edward R. 196
British Petroleum Company 148
Broadman, John R. 201
Broderick, Anthony J. 239
Building Owners and Managers Association 201
Bullard, Clark 201
Bushel, Julia 163
Business Publishers Inc. 201
Butane-Propane News Inc. 179

Author Index

C

California. Dept. of Finance 201
California. State Lands Commission
179
Canadian Petroleum Association 179
Canadian Petroleum Institute 153
Cannon, James 206
Chamber of Commerce of the United
States 179
Charter House Group 201
Chase Manhattan Bank 148, 149,
179
Chemical Specialties Manufacturers
Association 180
Christy, Francis T. 212
Colorado. Geological Survey 157,
180, 196
Commission on European Communities
201
Commodity Research Bureau Inc.
201-2
Conference Board 161, 202
Connecticut. Dept. of Commerce.
Technical Services Division 202
Corpus Publishers Ltd. 202

D

Dallas Morning News 202
Darmstadter, J. 202-3
DeGolyer and MacNaughten 155
Delaware. State Planning Office
203
Doernberg, A. 203
Dunkerley, Joy 203

E

Economist Intelligence Unit 170, 180,
181
Edison Electric Institute 149,
161-62
Emmings, Steven D. 203
Energy Modeling Forum 157
Environment Information Center 196,
204
Exxon Corp. 204

F

Fariday, Edward K. 170
Financial Post 157, 181
Florida. Energy Data Center 163,
182
Florida, University of. Bureau of
Business and Economic Research
204
Folk, H. 204
Foster Associates 205
Franssen, Herman T. 205
Fremond, Felix 205
French, Alexander 233
Future Requirements Agency. Den-
ver Research Institute University
of Denver 170

G

G.A. Turner and Associates 149
Gas Appliance Manufacturers Asso-
ciation 170
Georgia, University of. College of
Business Administration. Divi-
sion of Research 205
Gooch, Jennie C. 170
Good, Barry C. 205
Gordian Associates 205
Government Institutes Inc. 205
Great Britain. Central Office of
Information Reference Division
206
Great Britain. Dept. of Energy
206
Great Britain. Dept. of Energy.
Economics and Statistics Division
206
Gulf Publishing Co. 206
Gunwaldsen, D. 157
Gustaferro, Joseph F. 206

H

Hagen, Arthur W. 196
Hannon, Bruce 201
Harvey, Andrew 163
Hawaii. Dept. of Planning and
Economic Development 206

Hendrickson, Thomas A.   206
Herendeen, Robert   201
Herman, Stewart   206
Hirst, Eric   165, 207
Hoskin, Robert A.   165
Housgaard, Olaf   206
Houthakker, Hendrik S.   182

I

Idaho University.   Center for Business
   and Research   207
Illinois.   Commerce Commission.
   Accounts and Finance   163
Illinois.   Dept. of Business and
   Economic Development   207
Illinois.   Dept. of Mines and
   Minerals   158
Illinois State Geological Survey   182
Indiana.   Geological Survey   182
Indiana.   State Planning Services
   Agency   207
Inter Avia   234
International Air Transport Association
   234
International Atomic Energy Agency
   152
International Energy Agency   182
International Oil Scouts Association
   182
International Research Group   207
Interstate Oil Compact Commission
   183
Ion, D.C.   207
Iowa.   Development Commission.
   Research Division   208

J

Jackson, Frederick R.   196
Jacobs, Kenneth M.   239
Jones, David C.   157

K

Kansas University.   Institute for Social
   and Environmental Studies   208
Kentucky.   Dept. of Commerce   208
Kerr & Co. Engineers   151
Kirby, Ronald E.   234

L

Larwood, Gary M.   159
Latin American Center   208
Liepins, G.D.   208
Louisiana.   Dept. of Conservation
   183
Louisiana.   Division of Natural
   Resources and Energy   163, 170,
   183
Louisiana.   Geological Oil and Gas
   Division   183
Lundberg Survey Inc.   183

M

McGraw-Hill Inc.   208
Maine.   State Development Office
   208
Margolis Industrial Services   208
Martz, C.W.   196
Maryland.   Dept. of Economic and
   Community Development   208
Massachusetts.   Dept. of Commerce
   and Development   208
Metals Week   209
Michigan.   Geology Division   183
Michigan State University Graduate
   School of Business Administration.
   Division of Research   208
Miller-Freeman Publications   158
Minnesota.   Dept. of Economic
   Development.   Research Division
   209
Minnesota.   State Planning Agency.
   Office of Local and Urban Affairs
   209
Mississippi State University.   College
   of Business and Industry   209
Missouri.   Dept. of Natural Resources.
   Division of Research and Technical
   Information   158
Missouri.   Energy Agency   209
Missouri, University of.   Extension
   Division   209
Mitchell, B.R.   209
Montana.   Energy Advisory Council
   209
Montana.   State Division of Research
   and Information Systems   209

Author Index

264

SUBJECT INDEX

The following is a subject index to the list of additional sources contained in section III. Alphabetization is letter by letter and references are to page numbers.

A

Air conditioning systems 221
Airlines, fuel consumption 237
Anthracite 159, 219
Anti-freeze sales 180
Asphalt 189
Automobiles
   cost of operation 243
   passenger fuel economy 235
Auto ownership 243
Aviation industry 226, 231

B

Barge and towing industry 232
Bituminous coal 159
   consumption 161
   price 161
   See also Coal
Bunker Oil and coal 190

C

Carbon Black 189, 219
Coal
   bargeline companies 158
   companies 158
   consumption 161
   employment 158, 161

finance 157
industry 158
industry exports 158
marketing 157
mines 157, 158
mining methods 158
ownership 158
prices 157, 202
production 157, 160, 161, 202
reserves 158, 161
reserves by company 158
reserves by state 158
resources 161
stocks and reserves 158
sulfur content 160
top producing companies 158
trade 159, 161
transportation 157, 158, 159, 160
utilization 157
world production 158
   See also Bituminous coal
Coke
   prices 202
   production 202
Coking coal, supply and demand 160

Subject Index

Subject Index

M

Manufacturing industry
    energy consumption 202
    fuel consumption 220
Mass transit systems 241
Mass transportation 232, 238, 240, 246
Merchant ships 238
Metallurgical coke, supply and demand 160
Mineral industry 219
Mineral processing 158
Mineral production 160
Mining 158
Mobile rigs 184
Motor carrier industry 232
Motor fuel consumption 245
Motor gasoline stocks 178
Motor vehicles 238

N

Naptha stocks 178
Natural gas
    capital requirements 169
    direct sales 171
    drilling activity 172
    European data 171
    exploration 190
    helium occurence 171
    imports and exports 173
    prices 173, 202
    production 183, 202
    production costs 188
    rate schedules 169
    sales and revenues 174
    sales and use 169
    transmission sales 172
    underground supply 173
    year end reserves 173
Nuclear power
    contract awards 175
    electricity generated 176
    exports 175
    industry employment 175
    licensing 175
    new construction 174
    new plants 175, 176
    outages 177
    plant performances 174
    plant shutdowns 174
    supply 176
    world market 174

O

Octane rings 189
Office buildings, operating costs 201
Offshore
    equipment industry 184
    tanker terminals 184
    wells 179, 181
Oil equipment
    companies 183
    industry 187
Oil fields 183
Oil rigs 184
Oil shale 226
Oil well contractors 180
Oil wells 177, 178, 182
Oil wells, abandonments 178, 194
OPAEC energy data 222
OPEC data 186, 222
OPEC profile 209
Organic wastes 226
Outer continental shelf 224, 229

P

Panama Canal 235
Para-transit 234
Peat 219
Petrodollars 180
Petroleum
    Barge shipments 189
    dealer statistics 178
    estimated reserves 193
    exports and imports 178
    prices 177, 182, 185, 194, 202
    production 178, 183, 202
    product price 192
    refining 178
    transportation 178, 179
    See also Petroleum industry
Petroleum industry 195
    Arab countries 179
    Bahrain 180

Subject Index

U

Undersea soil mechanics 184
Uranium 219
   commercial sales 175
   deliveries and commitments 176
   distribution 175
   drilling 175
   exploration activities 176
   industry employment 175
   prices 202
   production 175, 202
   reserves 175
Urban mass transportation. See
   Mass transportation
Utility appropriations 161
Utility investment 161

W

Wildcat wells 177, 182
Wind power 197, 198